59.95

S0-ATF-688

Climate Change, Biodiversity and Sustainability in the Americas

Impacts and Adaptations

Francisco Dallmeier, Adam Fenech,
Don MacIver and Robert Szaro

EDITORS

A Smithsonian Contribution to Knowledge

Published in cooperation with Rowman & Littlefield Publishers, Inc.

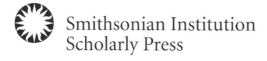

Smithsonian Institution
Scholarly Press

WASHINGTON, D.C.
2010

Colo. Christian Univ. Library
8787 W. Alameda Ave.
Lakewood, CO 80226

Published by SMITHSONIAN INSTITUTION SCHOLARLY PRESS
P.O. Box 37012, MRC 957
Washington, D.C. 20013-7012
www.scholarlypress.si.edu

In cooperation with
ROWMAN & LITTLEFIELD PUBLISHERS, INC.
A wholly owned subsidiary of The Rowman & Littlefield Publishing Group, Inc.
4501 Forbes Boulevard, Suite 200, Lanham, Maryland 20706
www.rowmanlittlefield.com

Estover Road
Plymouth PL6 7PY
United Kingdom

Copyright © 2010 by Smithsonian Institution

All rights reserved. No part of this publication may be reproduced, stored in a retrieval system, or transmitted in any form or by any means, electronic, mechanical, photocopying, recording, or otherwise, without the prior permission of the publisher.

British Library Cataloguing in Publication Information Available

Library of Congress Cataloging-in-Publication Data:
   Climate change, biodiversity and sustainability in the Americas : impacts and adaptations / Francisco Dallmeier ... [et al.], editors.
        p. cm.
     Includes bibliographical references and index.
     ISBN 978-0-9788460-7-7 (cloth : alk. paper)
     1. Biodiversity—America. 2. Climatic changes—Environmental aspects—America. I. Dallmeier, Francisco.
   QH101.C45 2010
   333.95097—dc22   2009040759

Printed in the United States of America

∞™ The paper used in this publication meets the minimum requirements of the American National Standard for Permanence of Paper for Printed Library Materials Z39.48–1992.

Contents

Preface

THE CHANGING climate is a significant driver of biodiversity and is already affecting many ecosystems throughout the Americas. It is necessary to mitigate and prevent these changes to preserve the biodiversity and ecological integrity of these regions. Increasingly, governments, organizations, industries, and communities need to consider adapting to the impacts of current and future changes in forest biodiversity in their planning, infrastructure and operations.

In order to begin addressing these issues, Environment Canada and the Smithsonian Institution's Center for Conservation Education and Sustainability co-hosted an international science symposium, "Climate Change and Biodiversity in the Americas" at the Smithsonian Tropical Research Institute (STRI) in Panama City, Panama, from 25 to 29 February 2008. The symposium brought together top researchers, industry representatives and managers of climate change and forest biodiversity research and monitoring activities from North, Central and South America and the Caribbean. It provided an opportunity for researchers and decision makers from a wide range of disciplines to share results and information in a pan-America event. The symposium program included invited keynote and plenary presentations, panel presentations, poster sessions, training sessions and study tours. Chairs of the symposium were Don MacIver, Director of the Adaptation and Impacts Research Division of Environment Canada in Toronto, Canada, and Francisco Dallmeier, Director of the Center for Conservation Education and Sustainability at the Smithsonian Institution, National Zoological Park in Washington, D.C.

The symposium was opened by STRI's director, Eldredge Bermingham, who stated that "having this Symposium in Panama has been especially significant, since the Isthmus has been cause and effect of major changes, both natural and anthropogenic. The scientific research conducted by the Smithsonian in Panama for almost 100 years allows us to provide a base of scientific knowledge to develop accurate measurements of environmental services to find the best management models for the biodiversity and biological richness of our planet in light of changes in the future."

Canada's Ambassador to Panama, Jose Herran-Lima, provided an impassioned speech identifying the main challenge facing science on the issue as the improvement of our understanding and predictive capacity of climate change and biodiversity with the goal to guide effective adaptive management and policy actions. Herran-Lima identified some cost-effective measures for doing this including:

- supporting biodiversity and climate monitoring networks (standardized protocols);
- prioritizing many small protected areas over a few large ones;
- protecting high priority conservation areas prior to development;
- managing the impacts from human activities;
- employing anticipatory adaptation and prevention;
- adopting a prevention approach to invasive species; and
- applying integrated mapping and modeling of landscapes and climate.

Herran-Lima identified the Panama statement that emerged from this symposium (see Appendix) as providing some good future directions. Specifically, he proposed that:

- international leadership interlinking climate change and biodiversity issues and needs must continue to be proactive;
- a leadership group is needed to implement guidance and advice from international conventions and agreements, as well as country-based strategies, science assessments and research activities; and
- integrated climate change and biodiversity actions are needed, such as:
  - establishing an integrated monitoring/modeling capacity of the synergies between climate, climate variability, climate change and biodiversity;
  - providing sound scientific expert advice to decision- and policymakers;
  - promoting standardized monitoring protocols and training capacities;
  - fostering community-based partnerships, including the integration of indigenous knowledge;
  - engaging education and outreach programs; and
  - investing financial investment in science and partnerships.

The keynote address was given by Thomas Lovejoy, president of the John Heinz Center for Science, Economics and the Environment, who gave an alarming overview of the threat of climate change to biodiversity in the Americas (Lovejoy and Karsh, 2008). He made it clear that biodiversity in the Americas is being transformed by human-induced climate change:

There have already been many noted changes among numerous species in times of nesting and flowering, as well as changes in geographical distributions, population dynamics and genetics. Increased CO_2 in the atmosphere has made oceans 0.1 pH unit more acid, negatively affecting tens of thousands of species that depend on calcium carbonate to build skeletons. On land, the alteration of the hydrological cycle has increased the probability of wildfire, which is devastating to biodiversity in regions with no previous adaptation to fire. Besides climate change, human activities are also accelerating loss in biodiversity.

Exotic species are being introduced far beyond their natural biogeographical boundaries. Native and non-native species alike must contend with pollutants for which they are unable to adapt. In the tropics, the widespread clearing and burning of forests is not only increasing CO_2 levels but also reducing biodiversity. Ironically, some of the destruction is due to the increased demand for ethanol and biodiesel (soybean and palm oil) by countries seeking to wean themselves off of oil. The habitats that do remain are becoming increasingly fragmented and isolated. Habitat fragmentation leads to genetic impoverishment and eventual extinction, as species can no longer adjust their ranges to climate change. Driven by habitat loss in tropical moist forests and by fragmented habitats and climate change, the current rate of extinction is 100 times faster than expected. If greenhouse gas emissions continue to run unchecked until 2050, future rates could be 1,000 times faster than expected. The impacts on biodiversity will be disastrous. Habitat fragmentation and climate change are the new challenges for biodiversity conservation. The current protected area system used in much of the Americas is insufficient given the realities of climate change. To ensure that biodiversity is protected—to offset synergistic interactions of fragmentation with other human effects—multiple large reserves are required, tens to hundreds of thousands of square kilometers in size, stratified along major environmental gradients to capture regional biota. All regional reserve networks and landscape connectivity must be wed with effective modeling of future climate change. To implement the changes necessary for sustainable ecosystems that are biologically healthy, functional and diverse, humanity also needs hope and the ability to dream of a glorious coexistence with a planet teeming with life. Part of the solution lies in the natural world and its ability to instill wonder. Awakening the biophilia inherent in humanity can improve the outlook for biodiversity if everyone has more contact with life on Earth and becomes more aware of the negative trends that threaten it.

The symposium was honored by other invited speakers representing international organizations including the World Meteorological Organization (WMO), the World Conservation Union (lUCN), the Convention on Biological Diversity (CBD), the UN Framework Convention on Climate Change (UNFCCC), the UN Education, Science and Cultural Organization (UNESCO), the World Wildlife Fund (WWF), Conservation International (CI), the Nature Conservancy, the Smithsonian Tropical Research Institute (STRI) as well as several universities and energy and forestry sector industries. Overall, there were 55 papers presented, 28 posters presented and over 130 scientists registered at the symposium. Twenty-one countries in the Americas were represented with speakers from Argentina, Belize, Bolivia, Brazil, Canada, Costa Rica, Cuba, Jamaica, Mexico, Panama, Peru, Puerto Rico, St. Vincent and the Grenadines, and the USA; posters presented from Guyana, Paraguay and Venezuela; and participants from Columbia, Guatemala, Haiti and Uruguay.

REFERENCES

Lovejoy, T. E., and M. B. Karsh. 2008. "Keynote Paper: Climate Change and Biodiversity in the Americas." In *Climate Change and Biodiversity in the Americas*, ed. A. Fenech, D. McIver, and F. Dallmeier, pp. 3–20. Toronto, Ontario, Canada: Environment Canada. 366 pp.

Acknowledgments

THE EDITORS wish to thank the following:

The officials who opened the symposium, including STRI Director Eldredge Bermingham, Canadian ambassador to Panama Jose Herran-Lima and, from ANAM, Panamá, Ligia Castro.

Thomas Lovejoy, president of the John Heinz Center for Science Economics and the Environment, who provided the keynote address.

The Symposium Science Advisory Committee including Robert W. Corell, the H. John Heinz III Center for Science, Economics and the Environment; Francisco Dallmeier, Smithsonian Institution, National Zoological Park; Ahmed Djoghlaf, Convention on Biological Diversity; William F. Laurance, Smithsonian Institution/INPA; Don MacIver, Adaptation and Impacts Research Division, Environment Canada; G. Arturo Sanchez-Azofeifa, Department of Earth and Atmospheric Sciences, University of Alberta; Holm Thiessen, InterAmerican Institute for Global Change Research; and Kenrick Leslie, Caribbean Community Climate Change Centre.

Adam Fenech as symposium coordinator, and chair of the symposium secretariat, which included Indra Fung Fook (northern administrator extraordinaire), Environment Canada; Audrey Smith (southern administrator extraordinaire), STRI; Marianne Karsh, Environment Canada; and Alfonso Alonso, Smithsonian Institution.

Trainers for SI/MAB Biodiversity Monitoring Plots (Cliff Drysdale, Alfonso Alonso, Alice Casselman and Amanda Munley) and Climate Modelling and Scenarios (Bill Gough, Neil Comer and Adam Fenech).

All speakers and all attendees.

All peer reviewers of the included papers, especially Tom Brydges.

Especially the Canadian International Development Agency for their funding of participants from less developed countries.

Other sponsors including Environment Canada; the Smithsonian Institution; the Smithsonian Tropical Research Institute; the H. John Heinz III Center for Science, Economics and the Environment; the World Meteorological Organization; International Union of Forest Research Organizations; the Convention on Biological Diversity; the Inter-American Institute for Global Change Research; the UNESCO MAB Programme; the Caribbean Community Climate Change Centre; ACER; and Arborvitae.

Introduction

Francisco Dallmeier,[1] Adam Fenech,[2] Don MacIver[2]
and Robert Szaro[3]

B IODIVERSITY—including species, natural communities and ecological systems— has evolved over time in response to the natural variation of the physical environment and biotic interactions within that environment to create a dynamic template that shapes how species interact and what species may (or may not) be able to persist in any given area (Landres et al., 1999; Parish et al., 2003; Szaro, 2008; Szaro and Williams, 2008). Confounding this natural dynamic template, human densities and activities impacting the physical environment have caused changes/losses in biodiversity at a more rapid rate during the past 50 years than at any time in human history and show no signs of yet diminishing (Millennium Ecosystem Assessment, 2005). The most important direct drivers of this loss and impacts on ecosystem services are massive landscape transformation such as urbanization and development, deforestation, natural resources extraction beyond the ability of species and habitats to bounce back, desertification, freshwater withdrawal from overland flows or groundwater and modification of water courses or water, massive coral reef bleaching, and damage to sea floors due to trawling and pollution, among others. Parallel to this unprecedented pace in global change are the synergistic effects of climate change, invasive alien species and infectious diseases (CCSP, 2009; Zaccagnini et al., this volume; Millennium Ecosystem Assessment, 2005).

For example, a study of six biodiversity-rich regions around the world representing 20% of the planet's land area found that 15% to 37% of all species in the regions considered could be driven extinct from the climate change that is likely to occur between now and 2050

[1] *Center for Conservation Education and Sustainability, Smithsonian Institution, National Zoological Park, P.O. Box 7012 MRC 705, Washington DC, 20013-7012 USA*

[2] *Adaptation and Impacts Research Division, Environment Canada, 4905 Dufferin Street, Toronto, Ontario, Canada M3H 5T4*

[3] *11585 Lake Newport Road, Reston, VA 20194 USA*

Corresponding author: Francisco Dallmeier (dallmeierf@si.edu).

(i.e., for mid-range climate warming scenarios) (Szaro, 2008; Thomas et al., 2004). Similarly, Wormworth and Mallon (2006) found that climate change affects bird species' behavior, ranges and population dynamics, with future climate change putting large numbers of bird species at risk of extinction. Estimates of these extinction rates range from 2% to 72% and depend on the region, climate scenario and ability of bird species to shift to new habitats.

Similarly, the present range and degree of protection of many plant species will rapidly erode as a result of climate change and the amount to which these losses can be compensated by the availability of newly suitable areas often affected by global change, the ability of the species to occupy new areas, and the degree to which we can influence changes in species distribution through management (Hannah et al., 2005). The rate of future climate and global change are likely to exceed the migration rates of most plant species (Nielson et al., 2005). The replacement of dominant species by new species may require decades, and extinctions may occur when plant species cannot migrate fast enough and suitable habitat is not available to escape the synergistic consequences of rapidly changing climate and human-induced global change.

Increasingly, international financial organizations, governments, industries and communities are considering higher standards and mechanisms for mitigating their impacts on climate and global change and adaptation mechanisms in their planning for development and operations. The impacts of climate change on biodiversity are of major concern to the Convention on Biological Diversity (CBD) as well as the work of the Intergovernmental Panel on Climate Change (IPCC) as it supports the United Nations' Framework Convention on Climate Change (UNFCCC). The risks to coral reefs and forest ecosystems have been highlighted, which has drawn attention to the serious impacts of loss of biodiversity on people's livelihoods. More recently, the CBD's Conference of the Parties has turned its attention to the potential impacts on biodiversity and ecosystems of the various options for mitigating or adapting to climate change and requested the Convention's Subsidiary Body on Scientific, Technical and Technological Advice (SBSTTA) develop scientific advice on these issues. The SBSTTA established an ad hoc technical expert group to carry out an assessment of the links between biodiversity and climate change—the results of which draw upon the best available scientific knowledge, including that provided by the Intergovernmental Panel on Climate Change (Djoghlaf, 2008). The report concludes that there are significant opportunities for mitigating climate change, and for adapting to climate change while enhancing the conservation of biodiversity. However, these synergies will not happen without a conscious attention to biodiversity concerns, nor without knowledge of likely future climatic changes.

This volume is the result of the symposium "Climate Change and Biodiversity in the Americas," convened 25–29 February 2008, in Panama City, Panama. The aim of the symposium was to provide a forum for leading scientists to present results of research and monitoring activities of climate change and forest biodiversity throughout the Americas. The goals of the symposium were to: review the baseline data and systematic observation networks to assess biodiversity conservation and policy responses to global climate change; integrate our knowledge of likely future changes on forest biodiversity as a result of a changing climate; report on predictive models and decision support tools to guide the design and selection of adaptation strategies from local to regional scales; and establish a framework for future collaborative research on climate change and biodiversity.

SPECIES RESPONSES TO CLIMATE CHANGE

Bird species responses to current trends in climate change differ both among species and among regions (Burton, 1995; Butler, 2003; Bunnell et al., this volume; Lehikoinen et al., 2004). Dramatic changes in response to climate have occurred in arrival dates, departure dates, over-wintering populations, spatial occupancy and relative density of those birds evaluated in British Columbia and elsewhere (Bunnell et al., this volume). Individual species impacts are further compounded by indirect climate-affected disturbances. For example, the passage of Hurricane Georges through Maricao, Puerto Rico, resulted in a species typical of open lowland forests, the White-winged Dove (*Zenaida asiatica*), being first reported in the forest after the hurricane when new microclimatic conditions facilitated its colonization (Tossas, this volume). Meanwhile, the Ruddy Quail-Dove (*Geotrygon montana*), which requires dense and close canopy coverage, was not reported again in post-hurricane surveys (Tossas, this volume).

Phenological phases of the endemic cactus *Mammillaria matlidae* were recorded as part of a long-term monitoring program for natural protected areas in a deciduous tropical forest near Querétaro City, Mexico (García-Rubio and Malda-Barrera, this volume). During the past four years, Querétaro City has experienced the most extreme rainfalls in 70 years including 115 mm in one day. Annually rainfall has ranged from 1,018 mm to 404 mm (García-Rubio and Malda-Barrera, this volume). These extreme events have helped to change our understanding of their effects on the vegetation phenology associated with rare climate events (García-Rubio and Malda-Barrera, this volume). This study illustrates that annual rain distribution influences seedling recruitment affecting population dynamics of *M. mathildae*. Such observations contribute to a better understanding of these relations and therefore facilitate improvement of the construction of prediction models to estimate species potential distribution after any climate event (García-Rubio and Malda-Barrera, this volume).

ECOSYSTEM RESPONSES TO CLIMATE CHANGE

Rapidly changing climate has potentially profound implications for biodiversity and has already become a significant driver of biodiversity affecting many ecosystems throughout the Americas (Lovejoy and Karsh, 2008). While these processes impact ecosystem structure and function individually, climate and land use are also interrelated, thus exacerbating the effects Zaccagnini et al., this volume). Climate extremes have been impacting biodiversity in the region with greater frequency, duration and severity than ever previously recorded (Karsh and MacIver, this volume; Roots, 2008). A number of studies have detailed climate and global changes and other associated disturbances, so have documented these impacts (Burkett at al., 2008; Chen et al., 2008; Knights and Joslyn, 2008; Lugo, 2008). Changes in the frequency, intensity, extent and location of disturbance will affect whether, how, and at what rate existing ecosystems will be replaced by new plant and animal assemblages (IPCC, 2002; Karsh and MacIver, this volume; Zaccagnini et al., this volume). These changes will be reflected in phenological changes, shifts of species occurrences, and increases in disturbances such as longer droughts, more frequent wildfires, higher temperatures and more intense storms, hurricanes and precipitation events (CCSP, 2009; Karsh and MacIver, this volume).

As an example, general circulation model scenarios of future precipitation patterns are highly uncertain in terms of future patterns of snow depth and the timing of snowmelt (Loik et al., this volume). Because snowfall provides the majority of annual soil water recharge in many western high-elevation North American ecosystems, Loik et al. (this volume) tested hypotheses about the linkages of snow depth to soil water content, recruitment and plant diversity at the ecotone between the Great Basin Desert shrub-steppe and Sierra Nevada conifer forest. They found that species richness and diversity were lowest on decreased-depth snow plots. There was a shift from co-dominance of the Great Basin Desert shrubs *Artemisia tridentata* and *Purshia tridentata* on ambient and increased-depth snow plots, to dominance of *P. tridentata* on decreased-depth snow plots (Loik et al., this volume). Moreover, such shifts may alter the availability of recruitment sites for the Sierra Nevada conifer species *Pinus contorta* and *Pi. jeffreyi,* highlighting complex interactions between the spatial and temporal patterns of physical and biotic factors at this site. Results suggest that an increased frequency of El Niño Southern Oscillation (ENSO) events will not impact the diversity of the Great Basin Desert–Sierra Nevada ecotone as much as would a decrease of snow depth and earlier melt timing, a possibility envisioned by an increasing number of climate model scenarios (Loik et al., this volume).

The synergism induced in Amazonia through simplifying ecological structure from the complexity of forest to the simplicity of soybean fields or ranchland, the increased probability of human-set fires, and complex ecological interactions mediated by climate, for example, bacterial diseases (Pounds et al., 2006), takes us into unknown bioclimatic territory (Bush and Lovejoy, 2007). In addition, we now know that the hydrological cycle, which produces a significant fraction of Amazon rainfall, is also responsible for 40% of the rain south of the forest in Brazil and northern Argentina. With 20% of the Brazilian Amazon deforested (and considerable further areas modified in various ways), the point at which an irreversible drying trend is triggered cannot be far away (Bush and Lovejoy, 2007).

For example, bird species from tropical islands are highly susceptible to natural events that disturb their habitats because of their limited distributions and small population sizes (Myers et al., 2000; IPCC, 2001). A case in point was the effect of the passage of Hurricane Georges through Maricao, Puerto Rico, in shifting the abundance of resident bird populations from the stable pattern attained during more than 60 years without a hurricane (Tossas, this volume). Fluctuations in the relative abundance were still occurring eight years after the passage of the natural phenomena, with eight species still having not reached prior baseline levels (Tossas, this volume).

Understanding the long-term trends in population and community dynamics in response to changing climate will require baseline data that describes the distribution and abundance of species across various spatial and temporal scales (Zaccagnini et al., this volume). The establishment of regional-scale, long-term monitoring programs are a key component in providing the types of data upon which the status and trends of species can be determined (Zaccagnini et al., this volume). In 2002–2003 the Biodiversity Group at the Instituto Nacional de Tecnología Agropecuaria (INTA), Argentina, started a regional-scale, long-term program that focuses on monitoring indicators of avian biodiversity in central Argentina (Zaccagnini et al., this volume). Their results in Argentine pampas ecosystems suggest that overall species richness is higher in habitats dominated by native vegetation and lower in habitats that have been converted to crop monocultures. Evaluation of spatial trend data suggests that variation in bird population dynamics exists and that using large-scale drivers (climate, land use and vegetation productivity) as predictor variables, their model explained 76% of the variation in the bird species richness data (Zaccagnini et al., this volume).

ADAPTING TO CLIMATE CHANGE

Adaptation

Protecting key ecosystem features involves focusing management protections on structural characteristics, organisms, or areas that represent important "underpinnings" or "keystones" of the overall system (CCSP, 2008). Adaptation approaches ultimately contribute to resilience, whether at the scale of individual protected area units, or at the scale of regional/national systems (CCSP, 2008; Fenech and Liu, 2008). The integration of adaptation and mitigation actions within the context of climate extremes, biodiversity conservation and sustainable development all call for greater synergy in implementing the Convention on Biological Diversity (CBD) and other relevant multilateral environmental agreements (Karsh and MacIver, this volume). It is necessary to develop adaptation and mitigation strategies to preserve the biodiversity and ecological integrity of these regions. The projected climatic changes for the next century (IPCC, 2007) are faster and more profound than any experienced in the last 40,000 years (Bush et al., 2004), and probably in the last 100,000 years (Bush and Lovejoy, 2007).

Synergies in protecting biodiversity and protecting the climate through adaptation measures will be facilitated by proposed joint action on biodiversity and mitigation measures through international pilot measures to reduce greenhouse gas emissions from deforestation and adaptation to deal with the changing climate impacts (Karsh and MacIver, this volume). It is expected that biodiversity's natural ability to adapt to a rapidly changing climate, especially given other environmental pressures such as habitat loss and fragmentation, will be insufficient to stem the additional losses of biodiversity expected because of climate change (Karsh and MacIver, this volume). As a result, planned or directed adaptation activities are urgently needed now to slow the increased rates of future biodiversity losses (Karsh and MacIver, this volume). As an adaptation strategy, maintaining biodiversity allows ecosystems to provide goods and services to communities while societies learn to cope with climate change (Karsh and MacIver, this volume).

Mitigation and Carbon Storage

Increased concern for climate change coupled with a willingness of some developed countries to adopt mitigation strategies has raised the interest for tropical tree plantations as a possible option (Sarlo et al., this volume). After the Kyoto Protocol, the social and economic benefits of carbon sequestration have launched tropical tree plantations to the forefront of possible mitigation strategies (Bristow et al., 2006; Cyranoski, 2007; Fonseca et al., this volume; Ishwaran, 2008). The current developing carbon market is favorable to tropical plantations (Olschewski et al., 2005), however concerns exist over the effects of large-scale monoculture plantations on biodiversity (Cyranoski, 2007). Although initial tree C storage might be somewhat lower than in monocultures, native species plantations cannot only be a good mitigation strategy but can also entail significant biodiversity benefits. Sarlo et al. (this volume) found native species plantations support differing groups of soil fauna with individual tree species acting as building blocks for diversity (Sarlo et al., this volume).

Fonseca et al. (this volume) suggested in the context of implementing the Kyoto Protocol, it is necessary to develop reliable data for estimating the carbon capture capacity of forestry and agroforestry ecosystems in order to economically compensate stakeholders for the environmental services provided. In this regard, they found that biomass and stored carbon in *Vochysia guatemalensis* and *Hieronyma alchorneoides* plantations at the EARTH University, located in the Caribbean zone of Costa Rica increased with age (Fonseca et al., this volume). Forest plantations of *V. guatemalensis* and *H. alchorneoides* reached 166.2 and 201.8 tC ha^{-1} at 14 years of age, respectively compared with a total of 154.9 tC ha^{-1} stored by secondary forests at 18 years of age. The soil stored between 82.5–86.3% of the total carbon content (Fonseca et al., this volume).

CONCLUDING REMARKS

We hope that you enjoy this collection of eight data rich peer-reviewed papers representing some of the leading thinking on biodiversity and climate change. The companion volume "Climate Change and Biodiversity in the Americas" contains the policy related presentations from the symposium (Fenech et al., 2008). Together these volumes and the Panama statement (see Appendix), stand as a continuing contribution to our attempts collectively to build our adaptive capacity to climate change.

REFERENCES

Bristow, M., J. D. Nichols, and J. K. Vanclay. 2006. Growth and Species Interactions of *Eucalyptus pellita* in a Mixed and Monoculture Plantation in the Humid Tropics of North Queensland. *Forest Ecology and Management*, 233:193–194.

Burkett, V., L. Fernandez, R. Nicholls, and C. Woodroffe. 2008. "Climate Change Impacts on Coastal Biodiversity." In *Climate Change and Biodiversity in the Americas*, ed. A. D. Fenech, D. MacIver, and F. Dallmeier, pp. 167–194. Toronto, ON: Environment Canada.

Burton, J. 1995. *Birds and Climate Change*. London: Christopher Helm.

Bush, M., and T. Lovejoy. 2007. Amazonian Conservation: Pushing the Limits of Biogeographical Knowledge. *Journal of Biogeography*, 34:1291–1293.

Bush, M. B., M. R. Silman, and D. H. Urrego. 2004. 48,000 Years of Climate and Forest Change from a Biodiversity Hotspot. *Science*, 303:827–829.

Butler, C. J. 2003. The Disproportionate Effect of Global Warming on the Arrival Dates of Short-Distance Migratory Birds in North America. *Ibis*, 145:484–495.

Chen, A. A., M. Taylor, D. Farrell, A. Centella, and L. Walling. 2008. "Caribbean Climate Scenarios for the Caribbean: Limitations and Needs for Biodiversity Studies." In *Climate Change and Biodiversity in the Americas*, ed. A. D. Fenech, D. MacIver, and F. Dallmeier, pp. 209–230. Toronto, ON: Environment Canada.

[CCSP] U.S. Climate Change Science Program. 2008. Preliminary Review of Adaptation Options for Climate-Sensitive Ecosystems and Resources. A Report by the U.S. Climate Change Science Program and the Subcommittee on Global Change Research. Ed. S. H. Julius and J. M. West. Authors: J. S. Baron, B. Griffith, L. A. Joyce, P. Kareiva, B. D. Keller, M. A. Palmer, C. H. Peterson, and J. M. Scott. Washington, DC: U.S. Environmental Protection Agency.

———. 2009. Thresholds of Climate Change in Ecosystems. A Report by the U.S. Climate Change Science Program and the Subcommittee on Global Change Research. Ed. D. B. Fagre, C. W. Charles, C. D. Allen, C. Birkeland, F. S. Chapin III, P. M. Groffman, G. R. Guntenspergen, A. K. Knapp, A. D. McGuire, P. J. Mulholland, D. P. C. Peters, D. D. Roby, and G. Sugihara. Reston, VA: U.S. Geological Survey.

Cyranoski, D. 2007. Logging: The New Conservation. *Nature*, 455:608–610.

Djoghlaf, A. 2008. "Perspectives on Climate Change from the Convention on Biological Diversity." In *Climate Change and Biodiversity in the Americas*, ed. A. D. Fenech, D. MacIver, and F. Dallmeier, pp. 21–33. Toronto, ON: Environment Canada.

Fenech, A., and A. Liu. 2008. "Climate Change Adaptation through Learning (ATL): Using Past and Future Climate Extremes Science for Policy and Decision Making." In *Climate Change and Biodiversity in the Americas*, ed. A. D. Fenech, D. MacIver, and F. Dallmeier, pp. 257–275. Toronto, ON: Environment Canada.

Fenech, A., D. McIver, and F. Dallmeier (eds.). 2008. *Climate Change and Biodiversity in the Americas*. Toronto, ON: Environment Canada.

Hannah, L., G. Midgley, G. Hughes, and B. Bomhard. 2005. The View from the Cape: Extinction Risk, Protected Areas, and Climate Change. *BioScience*, 55(3):231–242.

[IPCC] Intergovernmental Panel on Climate Change. 2001. *Climate Change 2001: Impacts, Adaptation and Vulnerability*. Contribution of Working Group II to the Third Assessment Report of the IPCC. Cambridge, UK: Cambridge University Press.

———. 2002. *Climate Change and Biodiversity*. IPCC Technical Paper V. Geneva: IPCC.

———. 2007. *Climate Change 2007: Impacts, Adaptation and Vulnerability*. Contribution of Working Group II to the Fourth Assessment Report of the IPCC. Cambridge, UK: Cambridge University Press.

Ishwaran, N. 2008. "Climate Change and Biodiversity: Challenges for the World Network of UNESCO Biosphere Reserves." In *Climate Change and Biodiversity in the Americas*, ed. A. D. Fenech, D. MacIver, and F. Dallmeier, pp. 35–43. Toronto, ON: Environment Canada.

Knights, R. D., and O. R. F. Joslyn. 2008. "Climate Change and Biodiversity in St. Vincent and the Grenadines." In *Climate Change and Biodiversity in the Americas*, ed. A. D. Fenech, D. MacIver, and F. Dallmeier, pp. 73–100. Toronto, ON: Environment Canada.

Landres, P. B., P. Morgan, and F. J. Swanson. 1999. Overview of the Use of Natural Variability Concepts in Managing Ecological Systems. *Ecological Applications*, 9:1179–1188.

Lehikoinen, E., T. H. Sparks, and M. Zalakevicius. 2004. "Arrival and Departure Dates. Birds and Climate Change." In *Advances in Ecological Research*, Vol. 25, ed. A. P. Møller, W. Fiedler, and P. Berthold, pp. 1–31. New York: Elsevier Academic Press.

Lovejoy, T. E., and M. B. Karsh. 2008. "Climate Change and Biodiversity in the Americas." In *Climate Change and Biodiversity in the Americas*, ed. A. D. Fenech, D. MacIver, and F. Dallmeier, pp. 3–20. Toronto, ON: Environment Canada.

Lugo, A. E. 2008. "Novel Tropical Forests: The Natural Outcome of Climate and Land Cover Changes." In *Climate Change and Biodiversity in the Americas*, ed. A. D. Fenech, D. MacIver, and F. Dallmeier, pp. 135–166. Toronto, ON: Environment Canada.

Millennium Ecosystem Assessment. 2005. *Ecosystems and Human Well-Being: Biodiversity Synthesis*. Washington, DC: World Resources Institute.

Myers, N., R. A. Mittermeier, C. G. Mittermeier, G. A. B. da Fonseca, and J. Kent. 2000. Biodiversity Hotspots for Conservation Priorities. *Nature,* 403:853–858.

Neilson, R. P., L. F. Pitelka, A. M. Solomon, R. Nathan, G. F. Midgley, J. M. V. Fragoso, H. Lischke, and K. Thompson. 2005. Forecasting Regional to Global Plant Migration in Response to Climate Change. *BioScience,* 55(9):749–759.

Olschewski, R., P. C. Benítez, P. C., G. H. J. de Koning, and T. Schlichter. 2005. How Attractive Are Forest Carbon Sinks? Economic Insights into Supply and Demand of Certified Emission Reductions. *Journal of Forest Economics,* 11:77–94.

Parrish, J. D., D. P. Braun, and R. S. Unnasch. 2003. Are We Conserving What We Say We Are? Measuring Ecological Integrity within Protected Areas. *BioScience,* 53(9):851–860.

Pounds, J. A., M. R. Bustamante, L. A. Coloma, J. A. Consuegra, M. P. L. Fogden, P. N. Foster, E. La Marca, K. L. Masters, A. Merino-Viteri, R. Puschendorf, S. R. Ron, G. A. Sanchez-Azofeifa, C. J. Still, and B. E. Young. 2006. Widespread Amphibian Extinctions from Epidemic Disease Driven by Global Warming. *Nature,* 439:161–167.

Roots, F. R. 2008. "Climate Change and Biodiversity in High Latitudes and High Altitudes." In *Climate Change and Biodiversity in the Americas,* ed. A. D. Fenech, D. MacIver, and F. Dallmeier, pp. 45–71. Toronto, ON: Environment Canada.

Szaro, R. C. 2008. Endangered Species and Nature Conservation: Science Issues and Challenges. *Integrative Zoology,* 3:75–82.

Szaro, R.C., and B. K. Williams. 2008. "Climate Change: Environmental Effects and Management Adaptations." In *Climate Change and Biodiversity in the Americas,* ed. A. D. Fenech, D. MacIver, and F. Dallmeier, pp. 277–294. Toronto, ON: Environment Canada.

Thomas, C. D., A. Cameron, R. E. Green, M. Bakkenes, L. J. Beaumont, Y. C. Collingham, B. F. N. Erasmus, M. Ferreira de Siqueira, A. Grainger, L. Hannah, L. Hughes, B. Huntley, A. S. van Jaarsveld, G. F. Midgley, L. Miles, M. A. Ortega-Huerta, A. Townsend Peterson, O. L. Phillips, and S. E. Williams. 2004. Extinction Risk from Climate Change. *Nature,* 427:145–148.

Wormworth, J., and K. Mallon. 2006. Bird Species and Climate Change. WWF-Australia, Sydney. Available online at: http://www.climaterisk.com.au/wp-content/uploads/2006/CR_Report_Bird-SpeciesClimateChange.pdf.

Avian Response to Climate Change in British Columbia, Canada—Toward a General Model

Fred L. Bunnell,[1] Michael I. Preston[2] and Anthea C. M. Farr[1]

ABSTRACT: British Columbia hosts 292 regularly breeding bird species that can be grouped into five migratory classes: resident, partial migrants, short-distance, long-distance and very-long-distance migrants. Using basic natural history features, expected response of members of each class over the past 40 years were predicted and then tested. Arrival and departure dates were largely as predicted on the basis of temperature changes and ice melt, except that some long-distance migrants arrived earlier than expected. Most species have increased overwintering (very-long-distance migrants did not). All showed shifts in range extent or relative abundance northward with increasing temperature. Dates of clutch initiation also changed with temperature. The general model of expected responses proposed and tested may allow agencies sufficient time to evaluate mitigative actions. Examples are provided of how these predictions could guide management actions.

Keywords: *climate change, biodiversity, migration, birds*

[1] *Centre for Applied Conservation Research, University of British Columbia, Vancouver, BC V6T1Z4, Canada.*
[2] *Biodiversity Centre for Wildlife Studies, P.O. Box 32128, 3651 Shelbourne Street, Victoria, BC V8P5S2, Canada.*
Corresponding author: F. Bunnell (Fred.Bunnell@UBC.CA)

INTRODUCTION

Bird species are responding to current trends in climate change (Burton, 1995; Butler, 2003; Karsh and MacIver, this volume; Lehikoinen et al., 2004). Published responses associated with global warming differ both among species and among regions (Lehikoinen et al., 2004). There are 357 bird species that occur most years in British Columbia (312 breeding; 292 regularly), all of which could respond somewhat differently to climate change. A model of expected responses would assist conservation planning in two ways: (1) predict changes in jurisdictional stewardship responsibility (stewardship sensu Dunn et al., 1999 as used by Partners in Flight, Panjabi et al., 2005) and (2) allow agencies time to consider potential mitigative actions. Stewardship responsibility is the proportion of a species' global population or range that occurs within a jurisdiction and reflects the global responsibility for that species within that jurisdiction. Ranges shifting in response to climate change also could shift global responsibility. Attributes of a model projecting likely shifts and preliminary tests of predictions are presented. Elsewhere, responses against specific climate variables are evaluated (for example, Huggard and Bunnell, 2005a, 2005b). Here, the focus is on temporal patterns as tests of model predictions. Empirical temporal patterns permit a simple illustration of trends to planners and decision makers.

DATA AND METHODS

The focus of this study is on trends within bird populations. Table 1 summarizes patterns of climate change within the nine terrestrial ecoprovinces of British Columbia (BC) (Figure 1). Analyses of meteorological data in Europe suggest that local temperatures reflect relatively large regions (Heino, 1994; Sokolov et al., 1998); nonetheless BC's ecoprovinces show different temperature responses (Table 1), probably because the province is so topographically diverse. Analyses of climatic data for the province indicate that temperature and moisture stress are not following the same temporal trajectories (Table 1; BC MoWLAP, 2002). Those differences are beyond the scope of this paper but could have profound implications to wetland-associated species.

Although data for most bird species treated here span 120 years, the focus is on comparisons between two decades (1960s and 1990s) providing some examples of longer term data. Changes in climate have not been uniformly distributed through space, but recent trends in warming are generally apparent around the 1960s (Walther et al., 2002; Lehikoinen et al., 2004).

Given that the data were not collected within a sample design but include many opportunistic contributions from naturalists, variable opportunity and effort were corrected in two ways. First, for evaluations of temporal trends, the database was restricted to a single record for any location on any date. That is, one record was considered sufficient to confirm the presence of a species in a given area on a given date even if many records existed. This restriction avoided bias and pseudoreplication resulting from extensive or long-term study in any particular area (especially those focused on a single species that produced a large pulse of records dominating underlying trends). Second, where all data were used, as for measures of relative abundance, differential sampling effort was corrected using rarefaction curves (see Preston, 2004; Bunnell and Squires, 2005). Sample sizes addressing temporal trends are those

TABLE 1. Trends in climate data by ecoprovince in British Columbia. Ecoprovinces are illustrated in Figure 1. All values significant at p < 0.05. NT = no trend (significant tests possible). — = insufficient data for tests of significance. Summarized from BC Ministry of Water, Land and Air Protection (2002). All differences are positive other than ice-free dates, which are tending earlier.

| | Temperature[a] | | | | Average Annual Precip.[b] | Ice-free Date[c] |
| | Spring | | Winter | | | |
Ecoprovince	Max	Min	Max	Min		
Georgia Depression (GD)	NT	1.2	NT	0.9	NT	—
Southern Interior (SI)	1.2	1.2	NT	2.4	4	—
Southern Interior Mountains (SIM)	1.4	1.4	NT	1.6	3	—
Coast & Mountains (C & M)	NT	1.2	1.9	0.7	2	—
Central Interior (CI)	1.7	2.2	NT	2.6	2	–8
Sub-boreal Interior (SbI)	1.3	2.1	NT	2.2	—	–7
Boreal Plains (BP)	NT	2.4	NT	NT	—	—
Taiga Plains (TP)	2.8	NT	NT	NT	—	–5
Northern Boreal Mountains (NBM)	3.6	3.9	NT	NT	—	–9

[a] °C per century (1895–1995)
[b] % per decade (1929–1998)
[c] Days per decade (1945–1993)

after data reduction to eliminate pseudoreplication for a single location and date. Sample sizes addressing questions of abundance or range expansion included larger datasets and were corrected for differential sampling effort. All bird data are archived at the Biodiversity Centre for Wildlife Studies (www.wildlifebc.org).

Analytical methods differed with the question asked. Changes in arrival or departure date represent minima or maxima of a distribution through time. Minima and maxima are very sensitive to sampling effort and their sampling distribution is wide, complex to describe statistically and difficult to interpret. Arrival and departure dates were evaluated in two ways. Quantiles were used where large samples (>3,000) had been digitized. The X% quantile is the date by which X% of a year's ordered records have been made. Quantile regression was used to directly fit flexible trend lines over time to the quantiles of the seasonal distributions of records. Quantile regression uses the full distribution of y-axis values to estimate quantiles at each point on the x-axis, whereas standard regressions fit to the mean of the distribution of y-axis values at each specific point on the x-axis. Two values were extracted to indicate recent trends in arrivals (5% quantile) and departures (95% quantile). One is the current trend, the first derivative of the fitted slope at the last year with data (usually 2004). The second is the average trend since 1960. This is the fitted value in year 2002 minus the fitted value in year 1960, divided by 42 years. The post-1960 trend is probably more relevant biologically than the current trend, because the latter can be affected by short-term fluctuations in either

FIGURE 1. Ecoprovinces of British Columbia (adapted from Demarchi et al., 1990).

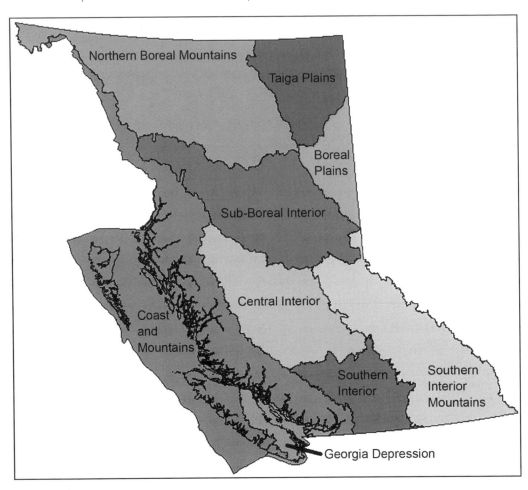

biological variables or simply in timing of observer effort. This approach was applied to Common Loon, Lewis's Woodpecker, Sandhill Crane, Surf Scoter, Swainson's Thrush, Yellow Warbler and Wilson's Phalarope (Table 2).

Where portions of the historical records were not yet digitized, or sample sizes were small, decadal differences in means of dates in the 5th and 95th percentiles for each decade (1960s and 1990s) were tested using parametric one-way analysis of variance (ANOVA) where variances were homogeneous and non-parametric Kruskal-Wallis tests where they were not. This approach was applied to Common Nighthawk, Heerman's Gull, and White-breasted Nuthatch (Table 2). Tests sometimes had to be geographically restricted to avoid overwintering or other potentially confounding variables. To assess changes in the probability of overwintering (December and January combined), logistic regression was used to evaluate presence: absence versus year for the period 1960–1999. A significant year effect indicated changes in overwintering. Data were restricted to areas of the province where overwintering currently occurs (Figure 2, see photospread).

TABLE 2. Tests of predictions of the general model of response for selected species in British Columbia. Shaded areas indicate where model predictions were met. NA indicates the test is not applicable to that species.

Group Species	Arrive Earlier[a]	Depart Later[a]	Overwinter[b]	Range Expansion[c]	Relative Density[d]
Resident					
White-breasted Nuthatch	NA	NA		YES***	i.d.
Partial					
Common Loon	−7.6***	+6.8***	YES**	NA	YES**
Surf Scoter	−4.3**	+0.5 ns	YES**	NA	YES ns
Short distance					
Lewis's Woodpecker	−0.5 ns	+1.6 ns	YES ns	NO	YES**
Sandhill Crane	−5.4***	+3.5**	YES**	YES***	i.d.
Long distance					
Heermann's Gull	−1.3 ns	+1.4*	YES ns	YES ns	YES*
Swainson's Thrush	−0.1**	+0.3 **	YES ns	YES ns	YES ns
Yellow Warbler	−0.1 ns	+0.9**	YES ns	YES*	YES*
Very long distance					
Common Nighthawk	+0.2 ns	+0.2 ns	none	YES ns	YES ns
Wilson's Phalarope	−0.8 ns	+0.9**	none	YES**	YES*

[a] days/decade

[b] year effect

[c] NA = not applicable (extended to 60° N in 1960)

[d] i.d. = insufficient data

*** p < 0.001 ** p < 0.01 * p < 0 .05 ns = nonsignificant

When evaluating changes in spatial occupancy or relative density, our primary interest was northward expansion. British Columbia spans 12° latitude (48–60°) North and 20° longitude (120–140°) West. Each 1° of latitude and 2° of longitude represents a 1:250,000 NTS (National Topographic Survey) cell, which is further divided into 16 1:50,000 NTS cells. Tests of spatial occupancy examined changes between the two decades in numbers of occupied cells within the 1:50,000 NTS cells. Tests of relative density evaluated effort-corrected density of observations within these same cells, using latitudinal belts to evaluate northward expansion. When testing, the distributions were constrained over latitude to the northernmost observation of either decade. Because the distribution of occupancy or relative density over latitude could assume a variety of distributions, the non-parametric Kolmogorov-Smirnov test was chosen to evaluate change. Although a northward expansion was anticipated, statistical tests were all two-tailed so shifts to the north or south could be evaluated. Cumulative frequencies of occupancy and relative density versus latitude were plotted (Figure 3) because that allows easy interpretation of the proportion of the population south or north of any given latitude in the two test decades (1960–1969 and 1990–1999). For example, about 60% of the species distribution or range for Lewis's Woodpecker

FIGURE 2. Changes in the probability of occurrence during winter (December and January occurrences only) in partial and long-distance migratory species in British Columbia. (a) Trumpeter Swan in southwest coast (48.00°–50.00° N latitude, 122.00°–128.00° W longitude) and southern interior (48.00°–50.00° N latitude, 116.00°–120.00° W longitude); (b) White-throated Sparrow on southwest coast (48.25°–48.30° N latitude, 123.00°–123.30° W longitude); (c) Turkey Vulture on southwest coast (48.00°–50.00° N latitude, 122.00°–126.00° W longitude); (d) Common Loon in southern interior (49.00°–52.00° N latitude, longitude variable). There were insufficient data to assess change at 51°N latitude (too few lakes). All trends, but Common Loon at 49°N are significant (p < 0.01).

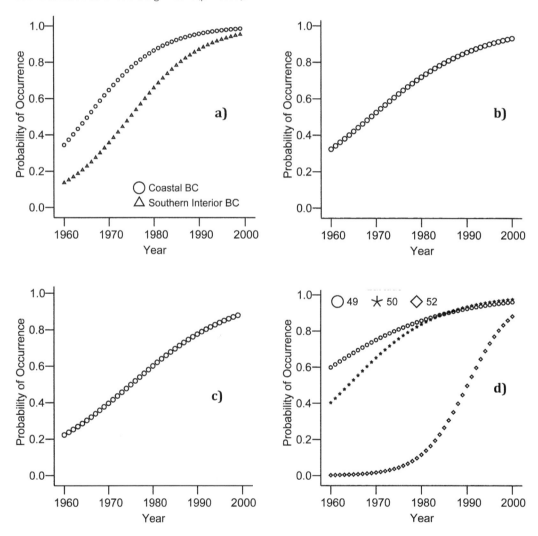

was south of 51° N in the 1960s, but only 40% was south of 51° N in the 1990s as the species' range extended northward (Figure 3).

Potential trends in clutch initiation were evaluated by quantile regression examining trends in early (5th percentile) and late (95th percentile) initiation. Potential changes in median initiation date with time were also evaluated using quantile regression.

GENERAL MODEL

Life history traits were used to make predictions of likely responses to climate change. The initial division of the general model uses migratory status of the species. Further subdivisions are based on features such as body size, food habits and breeding habitat. Based on these major features, predictions of species' response were developed and tested in terms of arrival date, departure date, (thus length of stay and potential changes in overwintering), range expansion, abundance and clutch initiation. Most correlates of body size (for example, multiple broods, second nesting) are not tested here. Reproductive responses are expected to be influenced by the phenological response of primary food sources during breeding (for example, piscivores, limited by ice on lakes, should show the strong responses in early arrival and subsequent reproduction).

FIGURE 3. Change in percentage distribution of occurrence (left) and relative abundance (right) of Lewis's Woodpecker (a and b) and Wilson's Phalarope (c and d) in British Columbia for the decades 1960–1969 and 1990–1999.

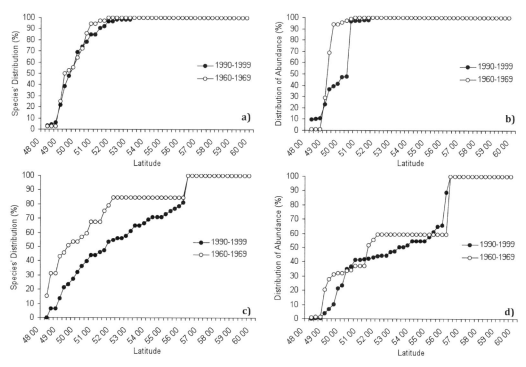

Resident Species

These are species that breed and winter on land (or over freshwater) within the province. There can be no change in arrival or departure dates, but range expansion northward within the province is expected. Currently, no range contractions due to climate is expected, because there is little tendency yet for the province to become drier or cooler (Table 1). The breeding season (clutch initiation) is expected to begin earlier. This group is least likely to show a mismatch with phenology of food resources because members exploit local cues. Because temperature minima are increasing faster than temperature maxima, frequent extension of the breeding season among species capable of multiple broods is expected. Species frequently producing two broods should show larger clutch size in the second brood than has been true historically.

Partial Migrants

Partial migrants are considered to be species that overwinter primarily in British Columbia, several at sea, but migrate to breeding areas within the province. Out of province migrants may augment numbers. Analyses of arrival and departure must be designed to keep overwintering individuals from obscuring trends.

   These species are constantly monitoring regional environments and should respond quickly to climate change. Generally, earlier arrival, later departure, increasing abundance of resident individuals (increasing stewardship responsibility), range expansion northward (increasing stewardship for species historically wintering primarily outside of BC), and earlier clutch initiation are expected. These species are unlikely to show a mismatch between breeding and the phenology of food resources because they are exploiting regional cues. For species with the capacity (generally smaller species), an extension of breeding season and the proportion of multiple broods should be seen. Such changes will appear as increased abundance and greater stewardship. Most larger species will be incapable of producing multiple broods, resulting in potential changes in relative abundance among large and small species. As with other species, responses are expected to be constrained by the phenological response of the primary food source during breeding and by body size.

Short-Distance Migrants

Short-distance migrants are defined as having some portion of the wintering population within 1,000 km of provincial boundaries (for example, about 38 °N). For this group, we expected earlier breeding and later departure (no great distance need be covered), range expansion northward creating increasing stewardship (especially among species whose habitats farther south may deteriorate; for example, increasing aridity should affect species seeking moist or mesic habitats), increasing numbers of resident birds within the province (especially among those species migrating shorter distances), earlier clutch initiation, and extension of the breeding season among species capable of multiple broods (thus increased abundance and subsequently greater stewardship) are expected. The potential for a mismatch between phenology and the food resource is increased over previous groups, because regional cues may not be accurate. Consequences of mismatch should appear in reproductive measures.

Among species whose nesting historically has been restricted to the north of the province, some may show increases in abundance as conditions in the Arctic change more than they do in British Columbia. That is, there is potential for the province to accumulate greater stewardship responsibility from the north as well as the south.

Long-Distance Migrants

Long-distance migrants are defined as having a significant portion of the wintering population >1,000 km distance, but <4,500 km (for example, 38 to 8°N, the southern end of Central America). For this group, changes along migration routes and stop-over points become increasingly influential. As well, the time and energy for molt prior to migration south may become a constraint. Departure times from winter and summer ranges are more likely to be under endogenous control or photoperiod than be influenced by climate variables (Berthold, 1996; Gwinner, 1986).

Earlier arrival dates among this group are not expected, but difficulties along the migration route could lead to later arrivals. No trend in departure dates was also expected, though if species arrive later they are also likely to depart later. The possibility of a mismatch between phenology of the food resources is increased because regional cues on wintering grounds may misrepresent conditions on the breeding grounds. Constraints noted earlier (food habits, body size, capacity for multiple broods) still apply, but may have less influence because time for molt and migration distance dominate. Combined effects could reduce survivorship.

The potential for range expansion in British Columbia remains considerable because many returnees from the wintering grounds are young birds with little affinity for natal areas. Among species whose nesting has been restricted to northern portions of the province, there is the potential for increases in abundance where conditions north of the province become less favourable than within the province in response to climate change. Wetter springs, for example, can reduce brood survivorship in all migratory classes. In such cases, southward shifts in range would increase provincial stewardship responsibility.

Very-Long-Distance Migrants

Very-long-distance migrants are defined as having a significant portion of the wintering population >4,500 km distance (for example, into South America). Within this group, changes along migration routes and stopovers become significant, as do the time and resources for pre-migratory molt. Relatively firm endogenous or photoperiod control over migration and possibly molt is expected.

Expectations are similar to those for long-distance migrants with the exception that the potential for mismatch with phenology of food resource is greater still, and there should be even less flexibility in departure dates. The potential for range expansion persists because many returnees again are young with relatively little affinity for natal areas. Again, species whose nesting has been restricted to north of British Columbia may show potential increases in abundance as conditions farther north deteriorate.

EXAMPLE SPECIES AND PREDICTIONS

For brevity, only selected tests of predictions are reported intended to span a range of natural history features and to illustrate information gained from testing. Table 2 provides an overview which is elaborated in sections following. Example species were selected to illustrate all migratory classes and a range of size classes, but were constrained by the degree to which data for individual species had been digitized.

Arrival and Departure Dates

Arrival and departure dates are relevant to only four of the five migratory groups (Table 2). Analyses were sometimes geographically restricted to eliminate confounding where overwintering appeared in data from later periods. Sufficient data were entered for seven species to permit estimation of long-term quantile regression: Common Loon (*Gavia immer*; n = 60,976; regression 1930 to 2004 for inland lakes), Lewis's Woodpecker (*Melanerpes lewis*; n = 7,628; regression 1890 to 2004 restricted to Southern Interior Mountains Ecoprovince), Surf Scoter (*Melanitta perspicillata*; n = 4,227; regression 1907 to 2004 for inland lakes), Sandhill Crane (*Grus canadensis*; n = 4,967; regression 1898 to 2004 for interior ecoprovinces), Swainson's Thrush (*Catharus ustulatus*; n = 18,898; regression 1890 to 2004 for central interior populations), Yellow Warbler (*Dendroica petechia*; n = 19,302; regression 1887 to 2004 province wide), Wilson's Phalarope (*Phalaropus tricolor*; n = 3,088; regression 1922 to 2004 province wide). Tests of decadal differences were limited to Heermann's Gull (*Larus heermanni*; n = 26) and Common Nighthawk (*Chordeiles minor*; n = 96).

Partial and short-distant migrants responded as predicted: by arriving earlier and departing later, some dramatically so. Of eight tests, all were in the appropriate direction, most significantly so (Table 2). The lack of significant change for Lewis's Woodpecker is likely caused by overwintering birds in a changing portion of the species' BC range. Long- and very-long-distance migrants also responded largely as predicted. Little change in arrival and departure dates was expected and, because of time for molt and the long migratory distance, little change in the duration of time spent by species within the province. Table 2 indicates that partial or short-distance migrants increased their residency time inland or within the province by 14.4 days per decade (Common Loon) or 8.9 days per decade (Sandhill Crane), while the change for very-long-distance migrants was 0.0 to −0.1 days per decade.

Analyses for larger databases sometimes showed differences across ecoprovinces (Table 3). For example, since 1960 Swainson's Thrush in most areas of the province is arriving earlier and departing later (Table 3). Interior and coastal populations of Wilson's Phalarope also appear to behave differently. Values in Table 3 are days/year over a 42-year period (see Data and Methods); to determine the total number of days and associated error, multiply by 42. For example, in the Central Interior Ecoprovince Swainson's Thrush was arriving 10 days earlier in 2002 (0.24 × 42) than in 1960, and staying 4.2 days later.

Overwintering

Winter was defined as December and January. Generally, the tendency for increased overwintering meant analyses of arrival and departure dates sometimes had to be limited to specific

TABLE 3. Trends[a] as days/year (SE in parentheses) from q = quantile regression analysis methods for arrival and departure times of Swainson's Thrush and Wilson's Phalarope in different ecoprovinces of British Columbia. See methods for explanation of current trend.

Species	Arrival		Departure	
Ecoprovince	**Current**	**Since 1960**	**Current**	**Since 1960**
Swainson's Thrush				
Southern Interior	*–0.25 (0.13)*	*–0.11 (0.08)*	**1.27 (0.29)**	**0.61 (0.20)**
Southern Interior Mountains	*0.18 (0.12)*	0.01 (0.08)	0.01 (0.58)	**0.84 (0.22)**
Central Interior	0.10 (0.34)	**–0.24 (0.11)**	–0.13 (0.62)	–0.10 (0.27)
Sub-boreal Interior	–0.01 (0.17)	*–0.10 (0.06)*	**1.30 (0.51)**	**1.08 (0.25)**
Northern Interior	–0.12 (0.15)	*–0.23 (0.12)*	**–1.85 (0.77)**	–0.70 (0.59)
Georgia Depression	–0.03 (0.20)	*–0.12 (0.10)*	0.24 (0.31)	**0.26 (0.12)**
South Mainland Coast	0.01 (0.44)	*–0.22 (0.20)*	–1.22 (0.89)	–0.21 (0.30)
Wilson's Phalarope				
Southern Interior	*–0.19 (0.13)*	*–0.16 (0.12)*	**1.61 (0.51)**	**0.75 (0.29)**
Georgia Depression	*0.18 (0.12)*	0.01 (0.11)	**0.63 (0.29)**	**1.07 (0.47)**

[a] Trend estimate 1–2 SEs from 0 are in italics; trend estimates >2 SEs from 0 are in bold.

regions of the province (for example, specific ecoprovinces of Figure 1). For example, records of overwintering Lewis's Woodpeckers in the Georgia Depression appear first in the 1920s; there are no winter records for the Southern Interior Mountains; in the valleys of the Southern and Central Interior Ecoprovinces, wintering records have been relatively consistent from the 1960s through the 1990s (Table 2). Analyses of arrival and departure dates for Lewis's Woodpecker were thus restricted to the Southern Interior Mountains.

Increased overwintering inland among partial and short-distance migrants was expected, but none among those traveling longer distances. Expectations were generally met for those migrating short distances. Lewis's Woodpecker on the coast is an exception, where fire suppression has eliminated favorable habitat. Neither the Common Loon nor Surf Scoter showed any increase in the northern interior ecoprovinces, but both species showed some tendency to overwinter inland in southernmost portions of the province, with overwintering increasing northward through time (for example, Figure 2d). No Sandhill Cranes were observed in the northeastern ecoprovinces during winter, but there was a slight increase in the number of winter period observations in the Southern Interior Ecoprovince. The lack of consecutive records throughout the winter months in any year suggests Sandhill Cranes may not overwinter in the southern interior, but may depart late in some years and return much earlier in others. A dramatic change occurred in coastal ecoprovinces. In 1979, 17 Sandhill Cranes were released near Vancouver in the Georgia Depression. There are only 40 winter observations recorded for coastal ecoprovinces prior to 1979. Of these, 26 occurred in a single NTS grid cell (Victoria) over 2 years, 1974 and 1975, neither of which appears anomalous in terms of weather. Conversely, the number of coastal winter records after 1979 is 430, 370 of these after 1995. Of these 430, 412 (96%) occurred in the Georgia depression near the introduction site. Winter observations in the southern interior may reflect birds moving between there and the coast.

The White-throated Sparrow, formerly a short-distance migrant, also shows increasing probabilities of overwintering (Figure 2). Prior to the 1960s most of the White-throated Sparrows breeding in the province were short-distance migrants from Washington; the probability of overwintering observations in 1960 was about 0.3. By 1999 the probability had increased significantly to about 0.9 (Figure 2b). The Trumpeter Swan illustrates the difficulties in discerning causes of the patterns observed. Most members of this species breed to the north of the overwintering areas in British Columbia (they are classed as short distance migrants between their breeding and winter range). The likelihood of overwintering birds has increased dramatically from the 1960s through the 1990s (Figure 2a). Whether this represents a simple increase in numbers or an interception of birds formerly migrating farther south to winter in Washington, Oregon, Montana, Idaho and elsewhere is unclear.

No species migrating very long distances showed a detectable tendency to overwinter; however, contrary to expectations, that was not true for long distance migrants (1,000–4,500 km; Table 2). Turkey Vultures breeding in British Columbia are reported to spend winter in Central and South America, and "leapfrog" over resident populations in California during migration (Kirk and Mossman, 1998). Their numbers in British Columbia during winter have increased significantly from the 1960s to the 1990s (Figure 2c). Similarly, when the apparently different populations of Swainson's Thrush are segregated by grouping ecoprovinces, there is no evidence of overwintering in the northeast, but an increase in overwintering in both the coastal and interior ecoprovinces. Using our definition of winter (December and January), Swainson's Thrush was observed on 2 days during winter prior to 1960 and 66 days after 1970. All winter observations were recorded in southern latitudes—about half were recorded in south coastal ecoprovinces and half in the southern interior ecoprovinces. The Yellow Warbler also showed unexpected overwintering. The first observation of Yellow Warblers in British Columbia during winter occurred in 1950, followed by 2 observations in the 1980s, and then 11 independent observations since 1990. The period post-1990 actually includes 90 observations, but many of these appear to be of the same bird. In all cases, the increased overwintering occurred in the southern portions of the province. For all putative long-distance migrants but Turkey Vulture, numbers of observations during winter are too few to permit statistical tests and the small increase in winter observations may not be biologically meaningful.

Range Expansion

Two measures of range expansion were evaluated: an increase in spatial extent northward and shifts in relative density within the area occupied. Only one resident species, White-breasted Nuthatch (*Sitta carolinensis*), is reported (Table 2). In the 1960s, the northernmost limit of its range was 51°15′ N; by the end of the 1990s the limit was 56°45′ N. Expansion of the range or relative density northward occurred across all migratory groups (Heermann's Gull still occupies a very limited area of the province, but has increased in abundance). Ranges for three of the species reported in Table 2 already extended to the northern boundary of British Columbia in the 1960s (Common Loon, Surf Scoter and Sandhill Crane). Increases in spatial extent of occupancy in the north were nonetheless apparent. For example, the Common Loon occupied more of the northernmost NTS cells in the 1990s. In the 1960s only 14% of occupied cells were north of 53°45′ N, whereas 34% were occupied in the 1990s. Because the species occupied the entire length of the province in both decades, the increase in size of the range is not significant (p = 0.107). The increase in relative density northward is significant (Table 2).

The response of Lewis's Woodpecker is insightful. There was a modest increase in occupancy of space north of 51° between the 1960s when less than 6% of occupied cells were north of 51° and the 1990s when more than 15% were north of 51° N (Figure 3a). That difference is not significant (Table 2) and the species remains concentrated in the southern part of the province. There has, however, been a significant shift of relative density to northern portions of this range. In the 1960s, <5% of observations occurred north of 50°; by the 1990s, >58% of observations were north of 50° N (Figure 3b). The marked shift in density within what is largely the same range is highly significant (Table 2). The implication is that Lewis's Woodpecker is much more constrained by other features of its environment than by climate. It is, in fact, largely limited to dry forest types. Our analyses suggest that its numbers are accumulating against the northern end of currently favorable vegetative habitat—particularly Ponderosa pine and Douglas fir.

Among species migrating longer distances it was acknowledged that range expansion could occur, but its consistency was surprising. By the 1990s the Swainson's Thrush expanded its range in central and northern areas of the province, though the bulk of numbers still remained concentrated in the south.

In the 1960s about 30% of observations of Swainson's Thrush were north of 54° N; by the 1990s that value had risen to about 50%. The shift, however, was not statistically significant. Yellow Warblers also had expanded their range into more northerly areas of the province by the 1990s, though the numbers remained concentrated in more southerly regions (Figure 4). In the 1960s, 76% of occupied NTS cells occurred south of 52° N; during the 1990s, <57% of occupied cells were south of 52° N. The northward range expansion was accompanied by a shift in relative density northward. In the 1960s the cumulative frequency distribution indicates that 97% of all effort-corrected observations were south of 51° N; in the 1990s, 90% of all observations were south of 51° N. Although slight, the expansion northward in relative density is significant (Table 2).

By the 1990s the northward extension of the range of Wilson's Phalarope, a very-long-distance migrant, had proceeded such that 33% of occupied cells were north of 54°15′ N, whereas in the 1960s the equivalent percentage was much further south at 51°30′ N (Figure 3c). That extension is significant (Table 2). The northward trend in relative density was even more pronounced (Figure 3d). In the 1960s about half of all observations were south of 52° N, but by the 1990s the southern half of error-corrected observations was south of 55° N. The difference between the distributions of relative density versus latitude is significant (Table 2).

Clutch Initiation

Where there were >40 records spanning the 2 test decades within an Ecoprovince (Figure 1), trends in clutch initiation were treated separately by Ecoprovince (Table 4). For Swainson's Thrush the tendency was for both median and late (95th quantile) nests to be initiated later. Late nests of Yellow Warblers also have been initiated later (Table 4). Both species are arriving

FIGURE 4. Distribution of effort-corrected relative abundance by NTS grid cell for Yellow Warbler in British Columbia during the 1960s (a) and 1990s (b). Squares: black = high, dark gray = medium, light gray = low, white = sampled, but no Yellow Warblers present. All other areas not sampled.

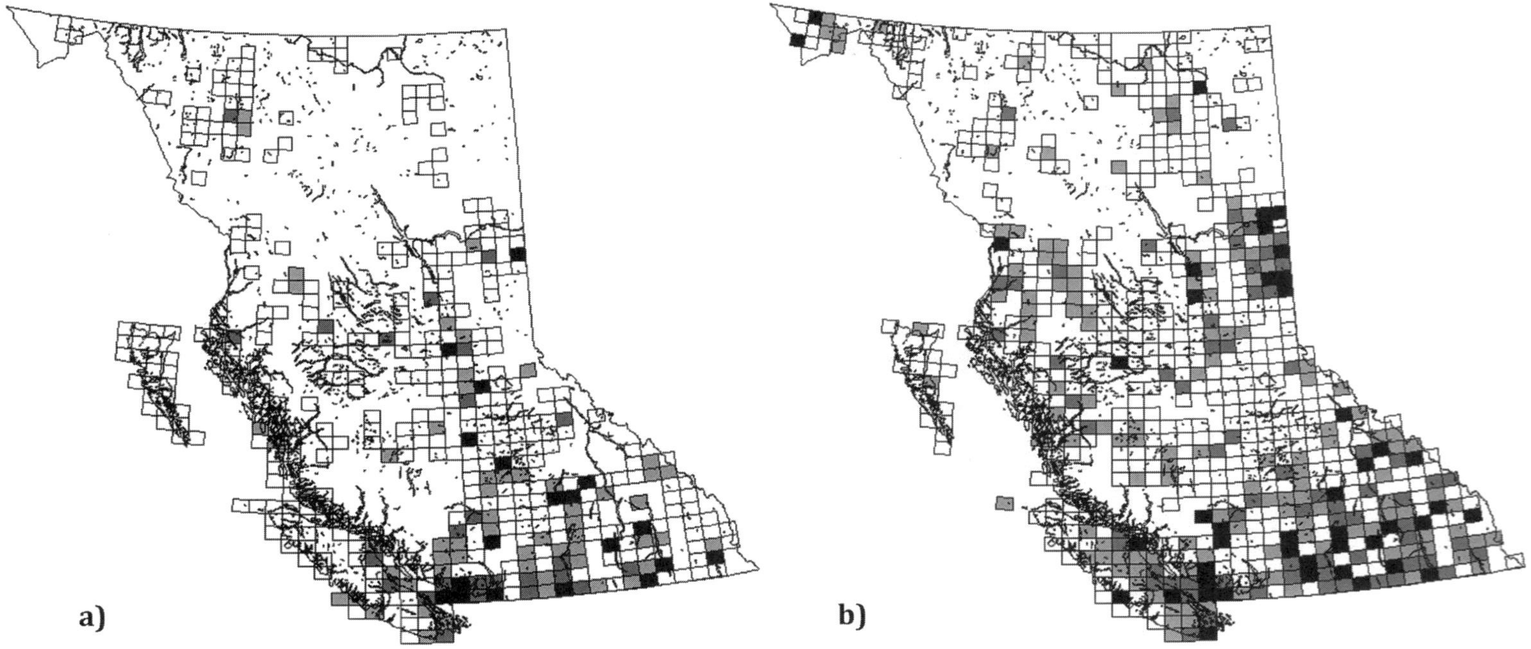

somewhat earlier, particularly in some ecoprovinces, and staying later (Table 2). The shift to later breeding may reflect greater flexibility in renesting attempts or a second nesting period. Values in Table 4 are days/year over a 42-year period (see Data and Methods); to determine the total number of days and associated error multiply by 42. For example, during the 42-year period 1960 to 2002, the apparent change in median date of clutch initiation for Swainson's Thrush is 42 times the value of Table 4, or 10.9 + 4.2 days.

DISCUSSION

Arrival and departure dates are evaluated by minima and maxima of the distribution of observations through time. These are very sensitive to sampling effort. Minima decrease and maxima increase with increasing effort, even if the underlying distribution remains exactly the same. This is particularly the case for distributions with long tails—like many migrating birds that start their migrations with a few early arrivals and end with a few stragglers. The sampling distribution of minima and maxima is wide and complex to describe statistically. There is little confidence in individual minima or maxima, but it is difficult to know exactly how imprecise these values are. Attempts to reduce these problems were made by employing quantiles; using the 5% quantile to indicate when arrivals started, while departures were indexed by the 95% quantiles. Statistically significant trends in these quantiles indicate that the trends in arrival and departure dates are real, rather than sampling artifacts.

Generally the simplistic a priori predictions made were met (Table 2), with some surprises. Spatial responses of the Lewis's Woodpecker, for example, suggest that existing long-lived vegetation (for example, preferred tree species) can dominate any effects of a shifting temperature envelope over the near term. Overwintering by long-distance migrants, such as Yellow Warbler and Swainson's Thrush, was not expected and suggests more future surprises in avian response to climate change. Marked regional differences in arrival and departure dates (for example, Swainson's Thrush and Wilson's Phalarope) suggest different but poorly understood migration routes into and out of the province. The pattern of response for Turkey Vulture suggests that its migration route is not fully understood.

TABLE 4. Trend[a] since 1960 (days/year) in early, median and late nesting dates for Swainson's Thrush and Yellow Warbler in three ecoprovinces of British Columbia (SE in parentheses). Positive values indicate later clutch initiation over time; negative values indicate earlier initiation.

Ecoprovince	Swainson's Thrush			Yellow Warbler		
	Early	**Median**	**Late**	**Early**	**Median**	**Late**
SI	0.07 (0.12)	**0.26 (0.10)**	**0.52 (0.22)**	–0.03 (0.10)	0.13 (0.24)	*0.41 (0.40)*
SIM	*–0.15 (0.24)*	*0.23 (0.16)*	**0.59 (0.16)**	0.02 (0.37)	0.22 (0.33)	0.26 (0.49)
GD	0.04 (0.09)	**0.28 (0.10)**	**0.80 (0.18)**	–0.10 (0.19)	0.10 (0.13)	**0.51 (0.18)**

[a] Trend estimate 1–2 SEs from 0 are in italics; trend estimates >2 SEs from 0 are in bold.

SI = Southern Interior, SIM = Southern Interior Mountains, GD = Georgia Depression (see Figure 1).

Although currently simplistic, the predictions offer some guidance to practitioners around stewardship sensu Dunn et al. (1999) and potential mitigation. Provincial stewardship clearly is increasing. Species are expanding their breeding or wintering ranges into the province (for example, White-throated Sparrow, White-breasted Nuthatch, Turkey Vulture). Partial and short distance migrants are increasing their stay inland or within the province (Table 2). The distribution of species breeding in British Columbia by migratory class is shown in Figure 5a. Of the 292 species breeding within the province, 183 (over 60%) fit tidily into one migratory class; the other half show mixtures (for example, a portion of the population is resident and a portion is short-distance migrants). Species were assigned to classes based on our understanding of what the majority of the population did. The exception is partial migrants, several of which are primarily resident (both breed and winter within the province) of which about half winter at sea.

Because of the inherent variation in movements within and among bird populations, considerable variability in response within migratory classes (Møller and Merila, 2004) is expected but the predicted pattern generally holds (Table 2). Of species breeding in British Columbia, partial migrants and short-distance migrants comprise 40 and 16 species, respectively. Most of these (about 20% of the breeding avifauna) are expanding their range within the province. This condition represents a marked increase in provincial stewardship that should be considered in conservation planning. The trend to increased stewardship is augmented by the expansion of the ranges of resident species within the province.

Current guidelines for mitigation of effects of climate change are likewise general. Wintering and breeding habitat typically differ. The increase in overwintering requirements, even among long-distance migrants (Table 2, Figure 2), indicates that more attention needs to be paid to wintering habitat. All partial and short-distance migrants tested showed increases in the likelihood of overwintering; appropriate habitats for these species are generally known. A shift in relative density and range expansion northward (where possible) was universal among species tested. Because mountain ranges in British Columbia are oriented north–south and perpendicular to prevailing winds, broad habitat types also are oriented north–south. This condition likely has facilitated northward expansion, with Lewis's Woodpecker providing an exception to the generality in that dry forest types have not yet expanded northward. Although bird species often make sharp distinctions between hardwoods and conifers when nesting or foraging, these are seldom restricted to specific tree species. Until climate change initiates shifts in the proportions of hardwoods and conifers, expansion northward of forest dwellers is likely to continue. In British Columbia this has implications to the approach to management, which focuses on sensitive species within specific administrative regions through the Identified Wildlife Management Strategy (BC MoE, 2004). The regional boundaries for the approach are no longer valid and are continuing to change.

Two other aspects of the general model, not fully treated here, also provide guidance to mitigation: foraging and nesting guilds. The model treats seven foraging guilds (Figure 5b). Among the breeding birds of the province, 31 and 38 fall into the piscivore and aquatic invertebrate foraging guilds, most of these on inland lakes and wetlands. Six nesting guilds are also recognized (Figure 5c). Birds using emergent vegetation as nest sites comprise about 10% of the breeding fauna (28 species). These specific foraging and nesting guilds are particularly susceptible to climate change. For example, species whose access to forage in deeper lakes has historically been impeded by ice, now show some of the earliest arrivals and latest

FIGURE 5. Primary migration (a), foraging (b) and nesting (c) strategies of 292 regularly breeding bird species in British Columbia.

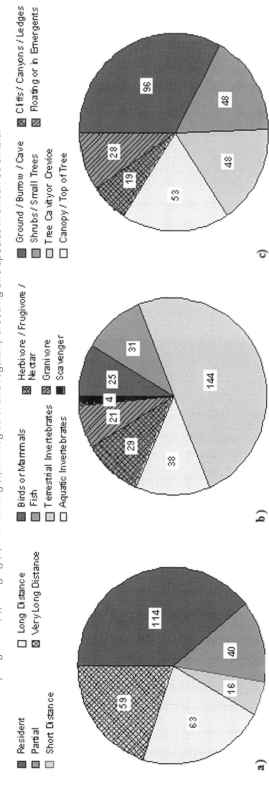

departures inland (Common Loon and Surf Scoter in Table 2). Species nesting on or in emergent vegetation are susceptible to changes in evapotranspiration. Emergent vegetation has a relatively shallow favorable rooting depth so favorable habitat could diminish quickly in convex, shallow ponds and wetlands as water levels decline. While still extensive, many of the province's wetlands are shallow. The likely response of bird species, coupled with bathymetric and climate data, could indicate where efforts at modified water management would achieve their greatest impact in sustaining biodiversity.

Nesting birds show strong affinities for broad vegetative types, such as grasslands, shrubs, hardwood trees and conifer trees. Coupling the general mode of avian response to models of climate change and anticipated changes in the physiognomy of ecosystems (for example, extension of grasslands, shifts between hardwoods and conifers) is a current extension of this study. Responses reported here indicate the promise of such combined models to guide mitigation of some effects of climate change.

CONCLUSION

Dramatic changes in response to climate have occurred in arrival dates, departure dates, overwintering populations, spatial occupancy and relative density of those birds evaluated in British Columbia. It is expected that the predictions will apply to significant portions of any group, rather than to all species within a group. Most of the responses reported here are as predicted by the general model. If the current promise is sustained, this tool will be available to assist management deliberations in the face of climate change, particularly when coupled with projections of evapotranspiration to assess vulnerable wetlands.

ACKNOWLEDGMENTS

Our work on climate change has been supported by the Forest Sciences Program of BC and the BC Ministry of Water, Land and Air Protection. We are thankful for the many naturalists in British Columbia that have reported their bird observations throughout the years. Much of the credit for collating and maintaining such a long-term database goes to R. Wayne Campbell.

REFERENCES

[BC MoE] British Columbia Ministry of Environment. 2004. Identified Wildlife Management Strategy. http://www.env.gov.bc.ca/wld/frpa/iwms/accounts.html.

[BC MoWLAP] British Columbia Ministry of Water, Land and Air Protection. 2002. Climate change website. http://wlapwww.gov.bc.ca/air/climate/indicat/maxmin_id1.html.

Berthold, P. 1996. *Control of Bird Migration*. London: Chapman and Hall.

Bunnell, F. L., and K. A. Squires. 2005. Evaluating Potential Influences of Climate Change on Historical Trends in Bird Species. Report to BC Ministry of Environment, Victoria, British Columbia.

Burton, J. 1995. *Birds and Climate Change*. London: Christopher Helm.

Butler, C. J. 2003. The Disproportionate Effect of Global Warming on the Arrival Dates of Short-Distance Migratory Birds in North America. *Ibis,* 145:484–495.

Demarchi, D. A., R. D. Marsh, A. P. Harcombe, and E. C. Lea. 1990. "The Environment." In *The Birds of British Columbia, Volume 1 (Nonpasserines: Introduction, Loons through Waterfowl)*, ed. R. W. Campbell, N. K. Dawe, I. McTaggart-Cowan, J. M. Cooper, G. W. Kaiser, and M. C. E. McNall, pp. 54–144. Victoria, British Columbia: Royal British Columbia Museum.

Dunn, E. H., D. T. J. Hussell, and D. A. Welsh. 1999. Priority-Setting Tool Applied to Canada's Landbirds Based on Concern and Responsibility for Species. *Conservation Biology,* 13:1404–1415.

Gwinner, E. 1986. "Circannual Rhythms." In *Endogenous Annual Clocks in the Organization of Seasonal Processes.* Berlin: Springer.

Heino, R, 1994. Climate in Finland during the Period of Meteorological Observations. *Finnish Meteorological Institution Contribution,* 12:1–209.

Huggard, D. J., and F. L. Bunnell. 2005a. Trends in Nesting Dates for Two Songbird Species in British Columbia. Report to BC Ministry of Environment, Victoria, British Columbia.

———. 2005b. Trends in Arrival and Departure Dates for Four Bird Species in British Columbia. Report to BC Ministry of Environment, Victoria, British Columbia.

Kirk, D. A., and M. J. Mossman. 1998. Turkey Vulture (*Cathartes aura*), The Birds of North America Online (A. Poole, Ed.). Ithaca, NY: Cornell Lab of Ornithology. Retrieved from *Birds of North America Online,* http://bna.birds.cornell.edu/bna/species/339.

Lehikoinen, E., T. H. Sparks, and M. Zalakevicius. 2004. "Arrival and Departure Dates. Birds and Climate Change." In *Advances in Ecological Research, Vol. 25,* ed. A. P. Møller, W. Fiedler, and P. Berthold, pp. 1–31. New York: Elsevier Academic Press.

Møller, A. P., and J. Merila. 2004. "Analysis and Interpretation of Long-Term Studies Investigating Responses to Climate Change, Birds and Climate Change." In *Advances in Ecological Research, Vol. 25*, ed. A. P. Møller, W. Fiedler, and P. Berthold, pp. 111–130. New York: Elsevier Academic Press.

Panjabi, A. O., E. H. Dunn, P. J. Blancher, W. C. Hunter, B. Altman., J. Bart, C. J. Beardmore, H. Berlanga, G. S. Butcher, S. K. Davis, D. W. Demarest, R. Dettmers, W. Easton, H. Gomez de Silva Garza., E. E. Iñigo-Elias, D. N. Pashley, C. J. Ralph, T. D. Rich, K. V. Rosenberg, C. M. Rustay, J. M. Ruth, J. S. Wendt, and T. C. Will. 2005. *The Partners in Flight Handbook on Species Assessment.* Version 2005. Partners in Flight Technical Series No. 3 [online], http://www.rmbo.org/pubs/downloads/Handbook2005.pdf.

Preston, M. I. 2004. A Little Effort Goes a Long Way. *Wildlife Afield,* 1:85–86.

Sokolov, L.V., M. Yu. Markovets, A. P. Shapoval, and Yu. G. Morozov. 1998. Long-Term Trends in the Timing of Spring Migration of Passerines on the Courish Spit of the Baltic Sea. *Avian Ecology and Behaviour,* 1:1–21.

Walther, G. R., E. Post, P. Convey, A. Merzel, C. Parmesan, T. J. C. Beebee, J.-M. Fromentin, O. Hoegh-Guldberg, and F. Bairlein. 2002. Ecological Responses to Recent Climate Change. *Nature,* 416:389–395.

Population Trends of Montane Birds in Southwestern Puerto Rico

Eight Years after the Passage of Hurricane Georges

Adrianne G. Tossas[1]

ABSTRACT: Puerto Rico, part of the Caribbean biodiversity hotspot, is frequently hit by hurricanes, which affect birds directly by decreasing their survival or indirectly when the resources on which they depend for foraging, roosting and nesting are destroyed. Georges, a moderate hurricane (category 3) according to the Saffir-Simpson scale of 5, crossed the island from east to west on 21–22 September 1998. It affected the forest structure of Maricao State Forest, a montane reserve in southwestern Puerto Rico spared by hurricanes for approximately 60 years, by destroying the canopy through defoliation and uprooting trees. The effects to resident bird abundance were assessed by comparing baseline point counts from 1998 with data from 1999, 2000 and 2007. Twenty-two species were detected in surveys throughout the length of the study, although the total number of species varied among years. A species typical of open lowland forests, the White-winged Dove (*Zenaida asiatica*) was first reported from the forest after the hurricane when new microclimatic conditions facilitated its colonization. Meanwhile, the Ruddy Quail-Dove (*Geotrygon montana*), which requires dense and close canopy coverage, was not reported again in post-hurricane surveys. Most species (15/17) showed declines in abundance in 1999, but half of those started showing gradual increases in 2000. In general, all species showed fluctuations in the mean number of individuals detected per count from

[1] *Department of Natural Sciences, University of Puerto Rico, Aguadilla, P.O. Box 6150, Aguadilla, Puerto Rico 00604-6150 (agtossas@gmail.com).*

year to year, but eight species, including initially common insectivores like the Puerto Rican (PR) Tody (*Todus mexicanus*), PR Woodpecker (*Melanerpes portoricensis*), and PR Vireo (*Vireo latimeri*), still had lower abundance in 2007 than prior to the hurricane. In contrast, species that were restricted to particular habitats within the forest, like the Adelaide's Warbler (*Dendroica adelaidae*), Antillean Euphonia (*Euphonia musica*) and Black-faced Grassquit (*Tiaris bicolor*), were able to expand their distribution after the hurricane. The resulting fluctuations in bird abundance, still occurring eight years after the hurricane, suggest a slow recovery of the forest structure or long-lasting damage to prey populations. Thus, it is recommended that long-term monitoring focus on how population trends relate to habitat changes and resource availability, particularly since more severe and frequent hurricanes may be expected with climate changes.

Keywords: *bird populations, Caribbean, habitat disturbance, Hurricane Georges, point counts, Puerto Rico*

INTRODUCTION

Bird species from tropical islands are highly susceptible to natural events that disturb their habitats because of their limited distributions and small population sizes (Karsh and MacIver, this volume; Myers et al., 2000; IPCC, 2001). Puerto Rico, part of the Caribbean biodiversity hotspot, provides habitat to 376 bird species (Puerto Rican Ornithological Society, unpublished data), including 17 endemics that are mostly found in the central mountain region. The island is frequently hit by hurricanes, which affect birds directly by decreasing their survival or indirectly when the resources on which they depend for foraging, roosting and nesting are destroyed (Wiley and Wunderle, 1993). Montane bird populations are at a major risk due to the stronger effects of wind gusts at higher elevations. In addition to natural events, resident birds are highly threatened by habitat destruction and competition with invasive species. Thus, long-term monitoring of montane bird species is necessary to keep track of their population dynamics and ensure their conservation, particularly since the frequency and intensity of hurricanes is expected to increase with ongoing climate changes (Webster et al., 2005).

Maricao State Forest (henceforth Maricao) is one of the largest and most diverse montane reserves in Puerto Rico. Seventy-three bird species have been reported from this site, including 13 of Puerto Rico's endemic species (Tossas and Delannoy, 2001). Populations of the endemics Elfin Woods Warbler and Puerto Rican (PR) Tanager (see Appendix for scientific names) are restricted to this reserve and few other montane areas. Maricao was struck by Hurricane Georges on 21–22 September 1998, after being spared by hurricanes since 1932 (Salivia, 1972; Scatena and Larsen, 1991; Boose et al., 2004). It was a moderate hurricane (category 3) according to the Saffir-Simpson scale of 5. However, its sustained winds of 184 kph and gusts of 240 kph (Bennett and Mojica, 1998) opened the canopy by severe defoliation, tree uprooting, and stem and branch breakage (Tossas, pers. observ.). Twenty-two months after the passage of Georges, 17 of 21 bird species had not recovered to pre-hurricane levels (Tossas, 2006). Maricao's avifauna showed stronger fluctuations than those reported from El Verde, part of the Luquillo Mountains in northeastern Puerto Rico, after the passage of Hurricane Hugo in 1989 (Waide, 1991; Wunderle, 1995).

Resident bird surveys were resumed in Maricao in 2007 to compare with results from 1998–2000 (Tossas, 2006). New point count records were compared with data collected previous to the passage of Hurricane Georges and a year and a half afterward. My main objectives were to determine (1) if bird populations have reached pre-hurricane levels eight and a half years after the natural event, and (2) if population dynamics were related to the species' diet and foraging level in the forest structure.

METHODS

Study Area

Maricao (18° 09′ N, 66° 59′ W) is in the westernmost part of the central mountain range of Puerto Rico. It comprises 4,150 ha with elevations ranging from 150 to 875 m (Silander et al., 1986). Annual rainfall and temperature from 1961 to 1990 averaged 232.6 cm and 21.7°C, respectively. The area contains subtropical moist forest, subtropical wet forest, and lower montane wet forest (Ewel and Whitmore, 1973). Dominant tree species include *Micropholis chrysophylloides, Terebraria resinosa, Linociera dominguensis, Homalium racemosum, Tabebuia schumanniana*, and *Eugenia stahlii* (Silander et al., 1986).

Point Count Surveys

The relative abundance of resident bird populations was surveyed with point counts before (1998) and after the passage of Hurricane Georges (1999, 2000, 2007). Point count stations were 100 m apart from each other, along two trails within the subtropical wet forest life zone. Both trails had similar floristic composition and vegetative structure. Fifteen point counts were conducted in each trail every year, from February to June, and only on days with fair weather. In each count, the observer stood in the center of the trail and, during 10 minutes, recorded all bird species seen or heard within a 25 m radius.

Statistical Analysis

Each species was classified as an insectivore, omnivore, nectarivore or frugivore, and as canopy or understory forager (see Table 1), following Waide (1996) and Wunderle (1995). The mean number of individuals detected in point counts, according to diet preferences and foraging level (species pooled), was compared with row x column tests of independence using a G-statistic. The mean number of individuals of each species detected in point counts was normalized with a logarithmic transformation (value x + 1) and compared among the four study-years with a one-way analysis of variance. A Mann-Whitney U test was used to determine the existence of significant differences in the relative abundance of bird species before (1998) and after (2007) the hurricane. Statistical tests followed Sokal and Rohlf (1981) and were performed in Statistix 8.0 (2003).

RESULTS

Twenty-two species of resident birds were detected in point counts throughout the length of this study, but the total number of species varied among study-years. The number of species declined in the two years following Hurricane Georges, but exceeded the pre-hurricane number in 2007 (Table 1). A new species, White-winged Dove, was first reported in Maricao after the hurricane. In contrast, a species reported in 1998, the Ruddy Quail-Dove, was not reported again in post-hurricane surveys.

The Bananaquit was the most abundant species in every study-year, but the proportion of individuals was different before and after the hurricane (Figure 1). While Bananaquits formed 17% of the total sample in 1998, they formed 27%–31% of the post-hurricane samples. The second, third and fourth most common species in 1998 (PR Tody, PR Tanager and PR Bullfinch, respectively), were again among the five most common species in post-hurricane years, although in different proportions (Figure 1). The PR Tanager and PR Bullfinch recovered their pre-hurricane levels by 2007, but not the PR Tody. This species formed 13% of the

FIGURE 1. Percentage of sample formed by the five most common bird species detected in point counts in Maricao State Forest (A) before (1998) and after the passage of Hurricane Georges: (B) 1999, (C) 2000, and (D) 2007. All other species detected were pooled in one category.

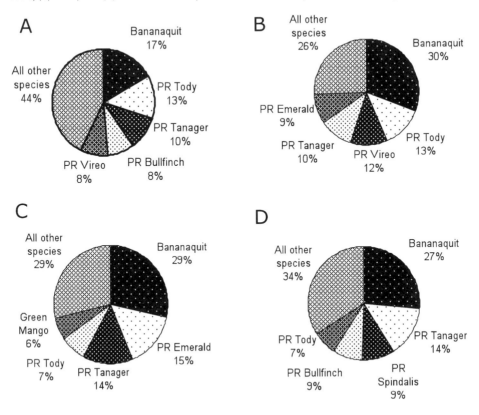

TABLE 1. Mean number of individuals recorded per point count ±SE, and proportion of sample (percentage within parentheses) of bird species in Maricao State Forest before (1998) and after Hurricane Georges (1999, 2000, 2007). The relative abundance of each species was compared among years with an analysis of variance. Species with significant differences (p < 0.05) in relative abundance are shown in bold.

Species	Diet, foraging level[a]	Mean ± SE (% of sample)			
		1998	1999	2000	2007
Scaly-naped Pigeon	F, C	0.23 ± 0.15 (4.5)	0.10 ± 0.07 (3.8)	0.07 ± 0.05 (2.0)	0.07 ± 0.05 (1.0)
White-winged Dove	F, C	0	0	0.03 ± 0.03 (1.0)	0.10 ± 0.06 (1.6)
Ruddy Quail-Dove	**F, U**	**0.17 ± 0.10 (3.2)**	**0**	**0**	**0**
Green Mango	N, U	0.10 ± 0.06 (1.9)	0.13 ± 0.06 (5.1)	0.20 ± 0.07 (6.1)	0.03 ± 0.03 (0.5)
Puerto Rican Emerald	N, U	0.27 ± 0.10 (5.2)	0.23 ± 0.09 (9.0)	0.50 ± 0.13 (15.3)	0.40 ± 0.11 (6.2)
Puerto Rican Tody	I, C	0.67 ± 0.14 (13.0)	0.33 ± 0.12 (12.8)	0.23 ± 0.09 (7.1)	0.43 ± 0.15 (6.7)
Puerto Rican Woodpecker	I, C	0.13 ± 0.08 (2.6)	0	0	0.03 ± 0.03 (0.5)
Puerto Rican Flycatcher	I, C	0	0	0	0.03 ± 0.03 (0.5)
Lesser Antillean Pewee	**I, U**	**0.23 ± 0.10 (4.5)**	**0**	**0.07 ± 0.05 (2.0)**	**0.03 ± 0.03 (0.5)**
Gray Kingbird	I, C	0.07 ± 0.05 (1.3)	0.07 ± 0.07 (2.6)	0.07 ± 0.07 (2.0)	0.03 ± 0.03 (0.5)
Puerto Rican Vireo	**I, C**	**0.43 ± 0.11 (8.4)**	**0.30 ± 0.09 (11.5)**	**0.07 ± 0.05 (2.0)**	**0.20 ± 0.07 (3.1)**
Black-whiskered Vireo	I, C	0.13 ± 0.06 (2.6)	0.07 ± 0.05 (2.6)	0	0.13 ± 0.06 (2.1)
Red-legged Thrush	**O, C**	**0.20 ± 0.09 (3.9)**	**0**	**0.07 ± 0.05 (2.0)**	**0.23 ± 0.08 (3.6)**
Pearly-eyed Thrasher	O, C	0	0.03 ± 0.03 (1.3)	0	0
Elfin Woods Warbler	I, C	0.30 ± 0.12 (5.8)	0.03 ± 0.03 (1.3)	0.13 ± 0.08 (4.1)	0.40 ± 0.14 (6.2)
Adelaide's Warbler	I, C	0.07 ± 0.05 (1.3)	0.03 ± 0.03 (1.3)	0.13 ± 0.10 (4.1)	0.17 ± 0.08 (2.6)
Bananaquit	**N, C**	**0.87 ± 0.12 (16.9)**	**0.80 ± 0.17 (30.8)**	**0.93 ± 0.20 (28.6)**	**1.73 ± 0.19 (26.9)**
Puerto Rican Tanager	**O, C**	**0.53 ± 0.13 (10.4)**	**0.27 ± 0.11 (10.3)**	**0.47 ± 0.14 (14.3)**	**0.93 ± 0.21 (14.5)**
Puerto Rican Spindalis	**F, C**	**0.30 ± 0.11 (5.8)**	**0.07 ± 0.05 (2.6)**	**0.07 ± 0.05 (2.0)**	**0.60 ± 0.14 (9.3)**
Antillean Euphonia	**F, C**	**0**	**0**	**0**	**0.13 ± 0.06 (2.1)**
Black-faced Grassquit	F, U	0	0	0.13 ± 0.09 (4.1)	0.20 ± 0.11 (3.1)
Puerto Rican Bullfinch	**F, C**	**0.43 ± 0.11 (8.4)**	**0.13 ± 0.06 (5.1)**	**0.10 ± 0.07 (3.1)**	**0.57 ± 0.16 (8.8)**
Total number of species		17	14	16	20
Total number of birds		154	78	98	194

[a] I = insectivore, O = omnivore, N = nectarivore, F = frugivore (fruit or seed), C = canopy, U = understory.

total sample in 1998, but only 7% of the 2007 sample. Similarly, the PRVI was the fifth most common species in 1998 (8% of total sample) but dropped to the ninth position in 2007, forming only 3% of the total sample.

Insectivores and frugivores formed the principal foraging guilds in Maricao, with nine and seven species, respectively. However, the nectarivorous and omnivorous guilds, consisting of three species each, dominated in abundance (Table 1, Figure 2). There were no statistical differences in the mean number of detections per count according to diet before and after the

FIGURE 2. Relative abundance of bird species in Maricao State Forest before (1998) and after (2007) Hurricane Georges according to (A) foraging guilds, and (B) foraging level.

A

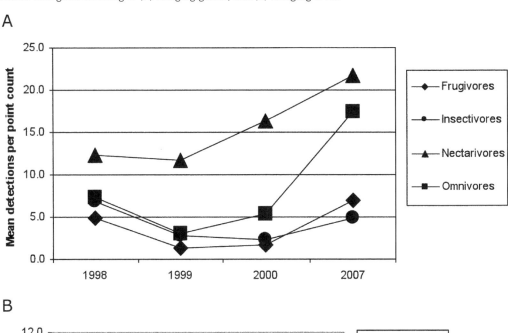

B

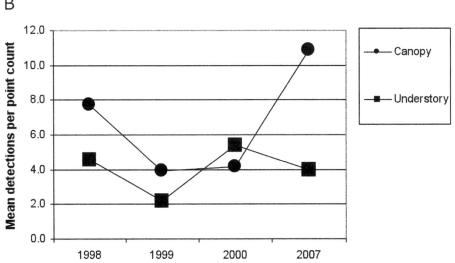

hurricane (G = 1.59, p = 0.66). Most species recorded in the surveys were canopy foragers (17/22), however, the preferred foraging level did not have a significant effect on mean detections per count before and after the hurricane (G = 0.26, p = 0.61).

All species showed fluctuations in the mean number of individuals detected per count from year to year, these variations being significant for nine species (Table 1). In a comparison of mean detections per point count before and after the hurricane, ten species had higher abundance in 2007 than in 1998, while eight had lower abundance. Four species had the same or similar abundance in both years. The abundance before and after the hurricane differed significantly for one species, the Bananaquit (Mann-Whitney U test, p = 0.003).

DISCUSSION

The main effect of the passage of Georges through Maricao was shifting the abundance that resident bird populations attained during more than 60 years without a hurricane. While in 1999 most species (15/17) showed declines in their abundance, more than half of those had already started showing gradual increases by 2000. However, fluctuations in the relative abundance were still occurring eight years after the passage of the natural phenomena, with eight species still having not reached baseline levels.

Only the abundance of four species—Black-whiskered Vireo, Red-legged Thrush, Puerto Rican Flycatcher, and Pearly-eyed Thrasher—resembled pre-hurricane levels by 2007. Nevertheless, the flycatcher and the thrasher are rare in the forest (Tossas and Delannoy, 2001), and were reported only in one study-year after the hurricane. Rare species may be able to better withstand the reduction in prey abundance in their home ranges by switching to less-damaged areas within the forest. According to Waide (1991), rare species in El Verde were also less affected by Hurricane Hugo than abundant species, demonstrating that population size is an important factor in determining how a species thrives after a natural disturbance.

Other species rarely recorded or absent from point counts previous to the hurricane due to particular ecological preferences—Adelaide's Warbler, Antillean Euphonia, and Black-faced Grassquit—showed increases in abundance by 2007. Initially, warblers were found mainly in the shrub vegetation of the drier south-facing slopes of the forest, and euphonias and grassquits were common in edges and openings tracking their preferred feeding resources (that is, mistletoes and grass seeds, respectively). However, warbler abundance doubled by the end of the study, and euphonias and grassquits, from being absent in baseline samples, together became 5% of the surveyed community.

The above-mentioned species probably benefited from microclimatic and vegetation structure changes that resulted from the opening of the canopy by defoliation and tree uprooting. Similarly, the White-winged Dove, a species common from open lowlands, was first reported in Maricao after the hurricane. This species has established with great success in the forest, occurring at present in higher amounts than the native columbids, Scaly-naped Pigeon and Ruddy Quail-Dove. In fact, by 2007 the pigeons still showed a threefold decline in their abundance, and only a few quail-doves have been distantly heard after the storm in 2006 and 2007, in dense vegetation at the base of south-facing ridges (C. A. Delannoy, pers. comm.; A. G. Tossas, pers. observ.).

Diet and preferred foraging level of bird species are other factors explaining the behavior of populations after a hurricane. According to results from El Verde, foraging guilds track the recovery of resources, thus nectarivore and frugivore abundance follows the appearance of flowers and fruits, as they become available after a hurricane (Waide, 1991). In the present study, bird abundance in all foraging guilds declined a year after Hurricane Georges, but nectarivores and omnivores showed a faster recovery, exceeding baseline levels by 2007. The nectarivorous guild was dominated by Bananaquits, the most common species in all study-years, while omnivores were mainly represented by the PR Tanager, which feeds mostly on fruits. The Bananaquit's resilience to changes in the forest structure and resource availability may have helped maintain a higher post-hurricane abundance for the last eight years. This result contrasts with the slow recovery of another nectarivore, the Green Mango, suggesting less adaptive behavior to habitat changes and not to lack of resources.

Both canopy and understory dwellers initially declined after the hurricane, but by 2007 only canopy species showed a marked increase in abundance, probably as a consequence of a higher number of species or displacement to a lower foraging strata. The post-hurricane abundance of understory dwellers, affected by the scarcity of two species requiring closed and dense forest, the Lesser Antillean Pewee and Ruddy Quail-Dove, was compensated by a rapid recovery of the PR Emerald.

The present study shows that some bird species in Maricao have not recovered eight years after the passage of Hurricane Georges, a moderate storm. While some species, like native columbids may still be affected from a long-lasting damage to the forest structure, another related effect may be low prey availability. The latter may explain why the insectivorous PR Tody, PR Woodpecker and PR Vireo, have not reached baseline levels by 2007. This finding contrasts with bird populations from El Verde, which showed signs of recovery a year or a year and a half after the passage of Hurricane Hugo (Waide, 1991; Wunderle, 1995). The slower recovery of Maricao forest and its birds (discussed in Tossas, 2006) may be related to the higher elevation of this study site compared to El Verde, stronger physical properties of Hurricane Georges, or adaptations of El Verde's flora and fauna to more frequent hurricanes crossing through this part of the island. Another probability is that forest growth in Maricao is constrained by unique conditions of its serpentine soils (Medina et al., 1994).

It is possible that fluctuations in Maricao bird populations reported in this study could have been caused by factors other than the hurricane, like individuals dispersing throughout the forest or beyond the borders of the reserve. In this scenario, population numbers may be related to differences in habitat quality or distribution of prey resources in the surrounding landscape. Thus, future studies should not only focus on population trends, but on how they relate to habitat changes and resource availability. In addition, long-term monitoring is needed to detect potential trends occurring as a consequence of more severe or frequent hurricanes, or other climate effects that can deteriorate habitat quality, like changes in temperature and precipitation.

ACKNOWLEDGMENTS

I am grateful to Luis A. Muñiz Campos, Beatriz Hernández and Ricardo Colón for assistance conducting the point count surveys. Funding for the initial surveys (1998–2000) came from a NSF grant (number HRD-9817642).

APPENDIX

Scientific Names of Bird Species Mentioned in the Text

Scaly-naped Pigeon (*Patagioenas squamosa*)

White-winged Dove (*Zenaida asiatica*)

Ruddy Quail-Dove (*Geotrygon montana*)

Green Mango (*Anthracothorax viridis*)

Puerto Rican Emerald (*Chlorostilbon maugaeus*)

Puerto Rican Tody (*Todus mexicanus*)

Puerto Rican Woodpecker (*Melanerpes portoricensis*)

Puerto Rican Pewee (*Contopus portoricensis*)

Gray Kingbird (*Tyrannus dominicensis*)

Puerto Rican Vireo (*Vireo latimeri*)

Black-whiskered Vireo (*Vireo altiloquus*)

Red-legged Thrush (*Turdus plumbeus*)

Pearly-eyed Thrasher (*Margarops fuscatus*)

Elfin Woods Warbler (*Dendroica angelae*)

Adelaide's Warbler (*Dendroica adelaidae*)

Bananaquit (*Coereba flaveola*)

Puerto Rican Tanager (*Nesospingus speculiferus*)

Puerto Rican Spindalis (*Spindalis portoricensis*)

Antillean Euphonia (*Euphonia musica*)

Black-faced Grassquit (*Tiaris bicolor*)

Puerto Rican Bullfinch (*Loxigilla portoricensis*)

REFERENCES

Bennett, S. P., and R. Mojica. 1998. *Hurricane Georges Preliminary Storm Report*. San Juan, Puerto Rico: National Weather Service.

Boose, E. R., M. I. Serrano, and D. R. Foster. 2004. Landscape and Regional Impacts of Hurricanes in Puerto Rico. *Ecological Monographs*, 74:335–352.

Ewel, J. J., and J. L. Whitmore. 1973. *The Ecological Life Zones of Puerto Rico and the U.S. Virgin Islands*. Research Paper ITF-18. USDA, Forest Service, Institute of Tropical Forestry, Río Piedras, Puerto Rico.

[IPCC] Intergovernmental Panel on Climate Change. 2001. *Climate Change 2001: Impacts, Adaptation and Vulnerability*. Contribution of Working Group II to the Third Assessment Report of the IPCC. Cambridge, UK: Cambridge University Press.

Medina, E., E. Cuevas, J. Figueroa, and A. E. Lugo. 1994. Mineral Content of Leaves from Trees Growing on Serpentine Soils under Contrasting Rainfall Regimes in Puerto Rico. *Plant and Soil*, 158:13–21.

Myers, N., R. A. Mittermeier, C. G. Mittermeier, G. A. B. da Fonseca, and J. Kent. 2000. Biodiversity Hotspots for Conservation Priorities. *Nature*, 403:853–858.

Salivia, L. A. 1972. Historia de los temporales de Puerto Rico y las Antillas, 1492–1970. Río Piedras, Puerto Rico: Editorial Edil, Universidad de Puerto Rico.

Scatena, F. N., and M. C. Larsen. 1991. Physical Aspects of Hurricane Hugo in Puerto Rico. *Biotropica* 23:317–323.

Silander, S., H. Gil de Rubio, M. Miranda, and M. Vasquez. 1986. Compendio Enciclopédico de los Recursos Naturales de Puerto Rico. *Los Bosques de Puerto Rico*, 10:210–236.

Sokal, R. R., and F. J. Rohlf. 1981. *Biometry: The Principles and Practice of Statistics in Biological Research*, 2nd ed. New York: W. H. Freeman and Co.

Statistix. 2003. *Statistix for Windows*. Version 8.0. Analytical Software, Tallahasee, FL.

Tossas, A. G. 2006. Effects of Hurricane Georges on the Resident Avifauna of Maricao State Forest, Southwestern Puerto Rico. *Caribbean Journal of Science*, 42:81–87.

Tossas, A. G., and C. A. Delannoy. 2001. Status, Abundance, and Distribution of Birds of Maricao State Forest, Puerto Rico. *El Pitirre*, 14:47–53.

Waide, R. B. 1991. The Effect of Hurricane Hugo on Bird Populations in the Luquillo Experimental Forest, Puerto Rico. *Biotropica,* 23:475–480.

———. 1996. "Birds." In *The Food Web of a Tropical Rain Forest*, eds. D. P. Reagan and R. B. Waide, pp. 364–398. Chicago: University of Chicago Press.

Webster, P. J., G. J. Holland, J. A. Curry, and H.-R. Chang. 2005. Changes in Tropical Cyclone Number, Duration, and Intensity in a Warming Environment. *Science,* 309:1844–1846.

Wiley, J. W., and J. M. Wunderle, Jr. 1993. The Effects of Hurricanes on Birds, with Special Reference to Caribbean Islands. *Bird Conservation International,* 3:319–349.

Wunderle, J. M., Jr. 1995. Responses of Bird Populations in a Puerto Rican Forest to Hurricane Hugo: The First 18 months. *Condor,* 97:879–896.

Regional Bird Monitoring as a Tool for Predicting the Effects of Land Use and Climate Change on Pampas Biodiversity

María Elena Zaccagnini,[1] Sonia Canavelli,[2]
Noelia Calamari[2] and Anne M. Schrag[3]

ABSTRACT: Large-scale processes, such as climate and land-use change, drive patterns of biodiversity worldwide. While these processes impact ecosystem structure and function individually, climate and land use are also interrelated, thus exacerbating their effects. Baseline data that describes the distribution and abundance of species across various spatial and temporal scales is necessary for understanding long-term trends in population and community dynamics. Regional-scale, long-term monitoring programs are a key component in providing the types of data upon which the status and trends of species can be determined. Birds are commonly chosen as indicator species in long-term monitoring programs due to their sensitivity to environmental change and migratory behavior, which results in local-scale changes having global-scale impacts. In 2002–2003, the Biodiversity Group at the Instituto Nacional de Tecnología Agropecuaria (INTA), Argentina, started a regional-scale, long-term program that focuses on monitoring indicators of avian biodiversity in central Argentina. The emphasis of this program is to examine the effects of land-use change and associated threats (for example, increased pesticide use) on metrics of avian populations, such

[1] *Instituto Nacional de Tecnología Agropecuaria (INTA), Instituto de Recursos Biológicos (IRB), De los Reseros y Las Cabañas S/N, 1712 Castelar, Argentina.*
[2] *INTA, Estación Experimental Agropecuaria Paraná, Ruta 11 Km 12.5, 3101 Oro Verde, Entre Ríos, Argentina.*
[3] *World Wildlife Fund, Northern Great Plains Program, 202 South Black, Suite 3, Bozeman, MT 59715, USA.*
Corresponding author: M. E. Zaccagnini (mzaccagnini@cnia.inta.gov.ar).

as bird species richness, composition, relative abundance and density. Preliminary evaluation of these data show spatial and temporal variations in landbird species richness related to land-use types. Therefore, birds could be used as indicators of change in these large-scale drivers and their interactive effects. Based on studies of the impacts of agriculture on birds in other regions of the world, such as Europe and North America, future changes in climate and land use to impact species richness and composition of avifauna in central Argentina might be expected. While conversion of land to agricultural uses in Latin American countries has been increasing, few long-term ecological monitoring programs are currently ongoing in Latin America. Thus, it is imperative to establish these programs as a tool for conservation in these countries.

Keywords: *monitoring, birds, land use, climate change, biodiversity, Argentina*

INTRODUCTION

Large-scale processes, such as climate and land-use change, drive patterns of biodiversity worldwide (Karsh and MacIver, this volume). While these processes impact ecosystem structure and function individually, climate and land use are also interrelated, thus exacerbating their effects (Dale, 1997; Szaro and Williams, 2008). Increasing demand for food and fuel due to a rapidly increasing world population is one of the primary factors leading to an unprecedented rate of land-use change, mainly due to the conversion of native vegetation to agricultural uses (Ojima et al., 1994). Such changes in land use and climate, and their accompanying ecological impacts, are important areas of concern for the conservation of biodiversity worldwide. These processes are progressing rapidly in countries that fall within the southern cone of Latin American (SCLA) (Argentina, Brazil, Paraguay, Bolivia and Uruguay), due to international demand for soybeans, especially in countries with rapidly increasing populations, such as China (Grau et al., 2005). Over the last several decades, but particularly since 1990, cereal and fuel crops have expanded rapidly in the region, largely through deforestation, replacement of crops and/or the conversion of lands previously used for extensive cattle raising (Zak et al., 2004; Thompson, 2007).

Concurrent with the expansion of agricultural production in the SCLA, agricultural intensification in Argentine pampas agroecosystems has also occurred, with increasing farm machinery sizes, large-scale agricultural labors, fertilizer and pesticides uses, among other technologies, all of which lead to more efficient farm practices (Panigatti et al., 2001; Viglizzo, 2001). Farmers are beginning to specialize in growing fewer crops, as opposed to their previously varied cropping techniques, which has led to a landscape dominated by monocultures and decreasing water-use efficiency and solar radiation (Caviglia et al., 2004). Additionally, agricultural practices associated with pest control, including herbicide and pesticide use, negatively impact water quality and increase hazardous effects on birds and other fauna (Hooper et al., 2002; Zaccagnini and Calamari, 2001). The amalgamation of those processes leads to noticeable habitat changes with detrimental effects on natural vegetation, decreases in diversity, and changes in composition and structure at regional and local scales (Solbrig, 1999). Changes in land use and climatic parameters are likely to continue in South American countries, particularly in Argentina, and both factors lead to biodiversity impacts.

Resident and migratory bird populations associated with farmlands have undergone significant declines and are experiencing range contractions in many parts of the world (O'Connors and Shrubb, 1986; Fuller et al., 1995). Loss of habitat and reduction in resource availability through land conversion is one of the causes, but climate change may also be responsible for these declines. Climate change may impact bird populations by reducing food resources and trophic interactions (Harrington et al., 1999), reducing breeding resources (with subsequent impacts on reproduction), earlier onset of egg laying (Stevenson and Bryant, 2000; Brown et al., 1999), and impacting species migrations (Gordo et al., 2005; Gordo and Sanz, 2006), distribution, abundance and composition across land-use gradients (Root, 1998, Winkler et al., 2002; Filloy and Bellocq, 2007).

Biodiversity monitoring has become a valuable tool for detecting the effects of changes in land use and climate. However, not all organisms or species possess the attributes that allow them to serve as good barometers of the effects of these large-scale changes. Birds are commonly used as indicator species in long-term monitoring programs due to their conspicuousness, use of a variety of habitats and resources, and sensitivity to environmental change and migratory behavior, which results in local-scale changes having global-scale impacts (Blair, 1999; Gregory et al., 2003). In addition, birds may respond to drivers that are operating at different scales (Wiens, 1989; Wiens et al., 2002). For example, climate operates at larger spatial and temporal scales, driving species distributions and migrations; land use and habitat availability operate at local and regional scales, thus influencing behavior, habitat use and habitat selection; and agricultural practices influence short-term and local-scale resource use (Clergeau, 1995). Consequently, a monitoring program needs to incorporate measurements at multiple scales and clearly delineate its limitations in order to provide useful information for land managers during the decision-making process.

In this paper we describe a regional-scale bird monitoring program that can function as a tool for predicting the impact of land-use and climate change at the eco-regional scale in Argentina. We propose the use of birds as indicators for orienting decisions in adaptation measures to adjust to those changes.

SCENARIOS OF CLIMATE CHANGE
IN ARGENTINE PAMPAS ECOSYSTEMS

The Argentine pampas has experienced changes in climatic parameters during the twentieth century. From 1916 to 1991 the central portion of the country experienced a trend of increasing precipitation (Barros et al., 1996). In some areas of the country, especially the northeast and central regions, precipitation and minimum temperature have shown increases while maximum temperatures decrease. Likewise, in the eastern and central portions of the country there has been a considerable increase in the frequency of extreme precipitation events, with a resultant increase in floods, destructive winds and hail associated with these events (República Argentina, 2007). Since the 1970s a clear displacement of the isohyets toward the western portion of the region has occurred, leading to a considerable increase in mean annual precipitation in the subhumid and semiarid zones (Hoffmann et al., 1997; Hurtado et al., 1996) (Figure 1a.) and increases in temporary and permanent flooding of agricultural fields (Magrin et al., 2005). These changes, together with others of cultural and technological origin, made possible the expansion of crop cultivation into the western edge of the previously humid region (Viglizzo, 2007). Models projecting future climate conditions for 2020–2040 predict a decrease in river volume within the Plata Basin due to increasing temperatures and subsequent increases in evaporation. In addition, models predict increasing hydric stress in northern and western parts of the country and continuing high frequency, intense precipitation events and floods in currently affected zones (Figure 1b). Although models projecting future climate change exist, uncertainty in predicted future changes hinder their ability to simulate the average and extreme precipitation in the La Plata River basin and neighboring zones. However, given the projections, natural vegetation and cultivated crops will likely experience changes due to shifts in precipitation, temperature and evapotranspiration, as well as impacts due to floods within the Argentine Pampas and Chaco region (2da.Comunicación Nacional de la República Argentina a la Convención de las Naciones Unidas sobre Cambio Climático, 2006). Consequently, these changes are likely to impact the distribution and abundance of species and communities.

BIRDS AS INDICATORS OF ENVIRONMENTAL CHANGE IN THE PAMPAS

Agricultural intensification has led to a widespread decline of biodiversity, particularly in farmland birds (Donald et al., 2001; Benton et al., 2003). This decline has been documented in Europe and the United States, where well-established, long-term monitoring programs exist and allow for time-series analysis of bird abundance and reproduction (Fuller et al., 1995; Siriwardena et al., 1998; Stephens et al., 2003). Declines in bird populations are also occurring in other parts of the world where agricultural expansion and intensification are homogenizing agricultural landscapes (Benton et al., 2003) and where highly toxic pesticides are used intensively or misused (Zaccagnini, 2006).

Birds can be used as "barometers" of environmental and climatic change at regional scales due to their dynamic use of ecosystems and agricultural landscapes throughout various ecosystems. Bird populations are very sensitive to changes in habitat availability, climate and resources (Harrington et al., 1999; Siriwardena et al., 2000; Murphy, 2003). Birds are also useful for tracking the toxicity of pesticides because of their broad food resource use and the consequent oral and dermal exposure in agroecosystems (Mineau, 2002, 2003). For these reasons, birds are commonly chosen as indicator species in long-term monitoring programs.

An example of local events having regional impacts emerged in 1995–1996, when massive mortalities of migratory Swainson's Hawks (*Buteo swainsoni*) in central Argentina occurred as the result of the misuse of a pesticide for controlling grasshoppers (Canavelli and Zaccagnini, 1996; Goldstein et al., 1996, 1999a, 1999b). The mortalities of that species and other non-migratory birds highlighted the detrimental effects of some agricultural practices

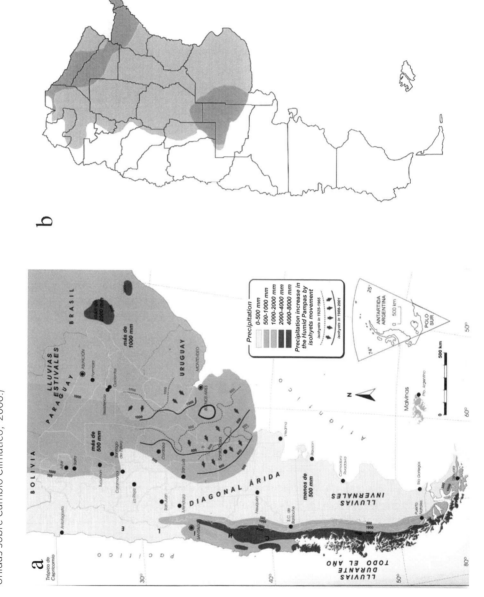

FIGURE 1. (a) Precipitation regimes of Argentina. Each line represents an isohyet of 500 mm to 4,000 mm (Naumann and Madariaga, 2003). (b) Areas experiencing increases in mean annual precipitation under predicted climatic scenarios. Darker areas indicate greater increases in annual precipitation. (Source: 2da. Comunicación Nacional de la República Argentina a la Convención de las Naciones Unidas sobre Cambio Climático, 2006.)

on birds in the most intensively used agricultural region in the country. It was suspected that the risks associated with pesticide use in the species' austral wintering grounds could lead to population declines (Woodbridge et al., 1995).

Consequently, in 2002 the Biodiversity Group at the Instituto Nacional de Tecnología Agropecuaria (INTA), with the support of United State Fish and Wildlife Service (USFWS) and the Neotropical Migratory Birds Conservation Act (NMBCA), started a regional-scale, long-term program that focuses on monitoring indicators of avian biodiversity in central Argentina. The emphasis of this program is to examine the effects of land-use change and associated threats (for example, increased pesticide use and habitat availability) on metrics of avian biodiversity, such as species richness, composition, relative abundance and density. However, in light of current predictions of global climate change, this regional, long-term monitoring program can serve as the basis for tracking the impact of climate conditions on bird diversity in Argentine grain-producing ecosystems.

REGIONAL BIRD MONITORING IN THE PAMPAS ECOSYSTEM

The success of biodiversity conservation and management decisions are dependent upon the availability of baseline data on species distribution and abundance across spatial and temporal scales (Noss, 1990). These data may be obtained by targeted monitoring or by omnibus surveillance monitoring. Focused monitoring may be a more efficient way of obtaining information for conservation planning and decision making, but surveillance monitoring may be an effective alternative when few data are available, such as in many developing countries (Nichols and Williams, 2006).

Few long-term ecological monitoring programs are currently ongoing in Latin America. Yet conversion of land in these countries has been increasing considerably as they anticipate leadership in the world grain economy (Grau et al., 2005). Within this context, the establishment of a bird monitoring program appears to be critical to track the effects of agricultural conversion on residential and migratory bird populations. The implementation of regional-scale, long-term monitoring programs is particularly important to ensure that high-quality information is being collected (Lovett et al., 2007) and that status and trends of species can be determined. The goal of our monitoring program in Argentine pampas is to obtain data on the status and trends of landbird populations at a regional scale, in order to analyze the effects of land use, pesticide hazards and climatic variables.

The monitoring program is based on point-count methods to survey species presence and distance sampling to estimate abundance and density of 20 focal species (Table 1) (Canavelli et al., 2003). Focal species include both resident and migratory species, as well as species from different trophic groups, such as resident raptors, that is, Crested Caracara (*Caracara plancus*), and the migratory Swainson's Hawk (*Buteo swainsoni*), and insectivorous resident passerines such us the White-browed Blackbird (*Sturnella supercilliaris*), and the migratory Fork-tailed Flycatcher (*Tyrannus savanna*); common granivorous species, for example, Eared Dove (*Zenaida auriculata*), which is highly mobile, and the resident Monk Parakeet (*Myiopsitta monachus*), which is not migratory, among other species (Canavelli et al., 2003). Most of the focal species have been affected by pesticides in previous years (Hooper et al., 2002, Zac-

TABLE 1. Focal species included in the regional bird monitoring survey in the pampas agroecosystems (alphabetically ordered by scientific name).

Spanish Common Name	Scientific Name	English Common Name
Lechucita de las vizcacheras	*Athene cunicularia*	Burrowing Owl
Taguató común	*Buteo magnirostris*	Roadside Hawk
Aguilucho langostero	*Buteo swainsoni*	Swainson's Hawk
Carpintero campestre	*Colaptes campestris*	Field Flicker
Paloma ala manchada	*Columba maculosa*	Spot-winged Pigeon
Paloma picazuro	*Columba picazuro*	Picazuro Pigeon
Paloma torcacita	*Columbina picui*	Picui Ground Dove
Milano blanco	*Elanus leucurus*	White-tailed Kite
Halcón plomizo o azulado	*Falco femoralis*	Aplomado Falcon
Halconcito común	*Falco sparverius*	American Kestrel
Chimango	*Milvago chimango*	Chimango Caracara
Tordo renegrido	*Molothrus bonariensis*	Shiny Cowbird
Cotorra común	*Myopsitta monachus*	Monk Parakeet
Perdiz chica	*Nothura maculosa*	Spotted Nothura
Cardenal copete colorado	*Paroaria coronata*	Red-crested Cardinal
Carancho	*Polyborus planctus*	Crested Caracara
Caracolero	*Rhostramus sociabilis*	Snail Kite
Pecho colorado	*Sturnella supercilliaris*	White-browed Blackbird
Tijereta	*Tyrannus savanna*	Fork-tailed Flycatcher
Paloma mediana o torcaza	*Zenaida auriculata*	Eared Dove

cagnini, 2006), giving an indication of their vulnerability to pesticide use within the ecosystem. For this reason, the 20 focal species will be used to serve as "indicators" of the impact of land-use practices on other species (Hutto and Young, 2002). Additionally, presence/absence information for other species in the area will allow analysis of land-use change impacts at the community level, using species richness or composition.

STRUCTURE OF THE REGIONAL BIRD MONITORING SURVEY IN THE PAMPAS

Since 2002, several 30 km transects have been surveyed annually in the central part of Argentina every January (austral breeding season) (Figure 2). The number of transects increased from 47 in 2003 and 2004, to 64 in 2005, to 72 in 2006, and to 84 in 2007 and 2008. As the area surveyed was expanded, the amount of land included increased to 255,000 km^2 with 212,000 ha of effective area surveyed on the last two years of the survey (2007 and 2008). Transects are located along unpaved secondary and tertiary roads at fixed locations. The survey area is stratified according to agrostatistical zones, defined by different land-use or production types and land-cover types. The survey began in 2002 with three zones representing agricultural, agri-livestock and forage-dairy production, and after 2004 the area was expanded to cover grasslands and pasturelands used for extensive livestock production but with some interspersed crops (Figure 2).

Transect locations were laid out according to a geographically stratified random design, by overlaying a 30 × 30 km grid over the area using ArcView 3.2 (ESRI, 1999). Thirteen strata were defined based on the agricultural zones and the provincial boundaries. Within each stratum, grid cells were selected in an systematic way with a random start, setting the number of cells to be surveyed in each agricultural zone to be proportional to the area of that zone. Within each cell, the route and the direction of coverage were randomly selected among all possible alternatives (Figure 2). Each route had 30 permanently marked points (total = 2,400 points). The first point in the route was randomly placed, with the others regularly spaced at intervals of 1,000 m to cover the majority of the grid cell and increase the probability of surveying different land-cover types. At each point, birds belonging to 20 focal species were surveyed using distance sampling (Table 1) (Buckland et al., 2001). Additionally, all birds seen or heard were recorded as present (Bibby et al., 2000). Observers used laser range-finders to measure distances from the observer to a single bird or the center of bird flocks, assuring that birds at point centers were not missed and accounting for bird movement by recording the initial locations of single birds or clusters, in order to meet the basic assumptions of point-transect methodology (Buckland et al., 2001). Counts were conducted between 0600–1,100 h and 1,500–2,000 h for logistical reasons, in order to cover the whole region in the minimum time. Land uses and habitat were described and their proportion estimated at each point of observation, in a radius of 200 m from the observation point. Focal bird densities are estimated with DISTANCE version 5.0 (Thomas et al., 2006; S. B. Canavelli, M. E. Zaccagnini, N. C. Calamari, and F. R. Rivera-Milán, unpublished manuscript).

Bird and land-cover estimates, together, allow for the calculation of various metrics of bird population dynamics, including species richness, composition, relative abundance and density. In the six years of bird monitoring in agroecosystems, the occurrence of a maximum of 173 species has been documented (166, 151, 155, 158, 173, 172) between 2002 and 2007, respectively. During the last two years (2006 and 2007) there was a considerable increase in the amount of area surveyed, which explained the increase in number of species observed. Maps of species richness (Image 1, see photospread) and relative abundance of 20 focal species were prepared by Canavelli et al. (2003) and Calamari et al. (2003–2007). Maps of relative abundance of migratory and resident focal species show interannual variations in spatial use by these species. For example, the migratory Swainson's Hawk (*Buteo swainsonii*) showed an interannual pattern of heterogeneous use of space (Image 2, see photospread), which may then be compared to the more homogeneous use by the resident Borrowing Owl (*Speotyto cunnicularia*) (Image 3, see photospread).

BIRD DIVERSITY IN THE PAMPAS IN RELATION TO LAND USE AND CLIMATE

Preliminary evaluation of data suggests that landbird diversity varies through space, in relation to land use (Image 4, see photospread). A model was developed to examine the impacts of large-scale drivers on bird species richness and composition at a regional scale (A. Schrag

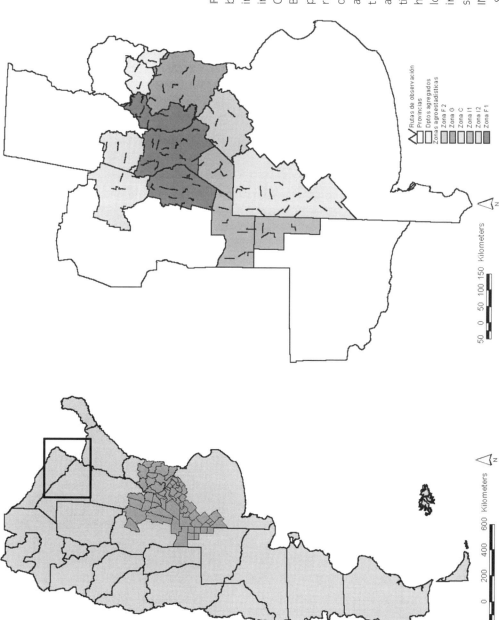

FIGURE 2. Area covered by regional bird monitoring in five Argentine Provinces (Entre Ríos, Santa Fe, Córdoba, La Pampa and Buenos Aires) within the pampas ecoregion. Lines represent routes for bird observation, and shaded areas differentiate agrostatistical zones (sampling area 255,000 km²; effective sampled area: 202,000 ha). Open rectangle shows location of future surveys in the Chaco ecoregion starting in 2008. (Source: INDEC, www.indec.mecon.gov.ar)

Rutas de observación
Provincias
Dptos agregados
Zonas agroestadísticas
Zona F 2
Zona G
Zona C
Zona I1
Zona I2
Zona F 1

N

50 0   50 100 150 Kilometers

200   0   200   400   600   Kilometers

N

et al. 2009). A generalized linear model (GLM) was built that describes bird species richness in relation to climate variables (maximum temperature, minimum temperature and annual precipitation), land-use variables (the percent cover of each of five land-use types, taken from the previously described monitoring program data) and the normalized difference vegetation index (NDVI—a measure of primary productivity and vegetation biomass). The impact of land-use gradients on species composition was then analyzed using canonical correspondence analysis (CCA), a constrained ordination technique.

Climate data were provided by the Instituto de Clima y Agua at the Instituto Nacional de Tecnología Agropecuaria and were derived from 20-year averages (1981–2000) for maximum temperature, minimum temperature and total annual precipitation from 39 climate stations that were located within the bounding box delineated by the following latitude and longitude: (−58.34, −30.82) and (−64.81, −36.07). This area encompasses all of the bird survey transects. Only stations that had a continuous record during the 20-year time period were used in the analysis. To associate broad-scale climatic gradients that may affect landbird habitat with bird survey transect locations, spatial interpolation of the pointbased climate station data was performed using the Universal Kriging tool in ArcGIS 9.1 (ESRI, 2005). This method allowed us to sample interpolated climate values for the location of the midpoint of each bird monitoring transect.

To determine whether species richness depends upon broad-scale variation in vegetation measures such as productivity, which may act as surrogates for habitat quality, the mean normalized difference vegetation index (NDVI) value was sampled for the location of each transect midpoint using satellite data. The NDVI was calculated for each transect midpoint by sampling a 3×3 pixel area (each pixel is 1 km^2) centered on the midpoint location and averaging the values of the nine resulting cells. Images were used beginning in August 2002 through July 2006 in order to encompass the austral summer and the accompanying previous spring. The NDVI values were compiled by season and averaged across the study period to produce one value for each transect for each season.

Evaluation of spatial trend data suggests that variation in bird population dynamics exists in this ecosystem. Using large-scale drivers (climate, land use and vegetation productivity) as predictor variables, the model explained 76% of the variation in the bird species richness data, suggesting that these large-scale ecosystem drivers indeed are impacting the spatial distribution of bird populations in this region. In addition, species richness was found to be positively correlated ($r = 0.77$) with cover of native vegetation and negatively correlated ($r = -0.61$) with cover of annual cultivars, such as soybeans. Furthermore, multivariate analysis of species composition along land-use gradients suggests that distinct groups of species are selecting optimal habitat along the existing land-use gradients (Schrag et al. 2009).

These results suggest that overall species richness is higher in habitats dominated by native vegetation and lower in habitats that have been converted to crop monocultures. Based on these results we conclude that birds could serve as effective indicators of changes in land use and that they are responding to large-scale drivers, such as climate and land-use change, at the regional scale of this study. Future changes in these drivers, and/or their interactive effects, may lead to overall changes in species richness and composition of avifauna in central Argentina. Increases in agricultural intensification in the region are expected to lead to decreases in overall species richness and shifts in composition toward agriculturally associated species.

FINAL COMMENTS

A long-term, regional-scale monitoring program, such as the one described here, provides essential baseline information for assessing the status and trends of birds as biodiversity indicators in agroecosystems. High-quality, long-term data series are essential for detecting the impacts of land use and climate change. The repeated monitoring described here will allow, in the long term, the information needed to analyze spatial-temporal trends of birds diversity, land uses and climate change.

Even taking into account the limitations of the "omnibus surveillance monitoring" for its potential use for many different purposes, it is important at least to consider this type of monitoring program in developing countries. However, it is also important to consider the level of investment required. First, it is necessary to be able to provide continuous training and capacity building to develop and sustain the monitoring system over time. Second, an institutional demand for quantitative data for adaptive management should be present. Third, knowledge gained should be transferable to the general public to make citizens aware of the importance of the program and the ability of the citizens to promote the development of policies to sustain such programs. Finally, these programs require guarantees of support and funding to secure long-term implementation, which is not an easy task in developing countries. But, if implemented, especially in addition to more intensive studies, such as focused monitoring on those species that are already endangered, this type of program may be an important tool for countries to provide information to address the status of the 2010 biodiversity indicators and, hopefully, contribute to a decision-making process that promotes biodiversity worldwide.

ACKNOWLEDGMENTS

This project has been possible through partial support from the Canadian Wildlife Service, and several grants received by U.S. Fish and Wildlife Service, Western Hemisphere International Office (1997–2002) and the Neotropical Migratory Birds Conservation Act (2003–2005), and Projects #745 and AERN 2622 from the Instituto Nacional de Tecnología Agropecuaria. We thank all participants who collaborated with us during the different stages of training, field survey and data entry. We extend special thanks to Dr. Frank Rivera, who trained and advised on distance sampling, and Dr. Rob Bennetts for his advice to Anne Schrag for portions of this work. We especially acknowledge the Instituto de Clima y Agua from INTA-CNIA Castelar, Carlos Di Bella and Pablo Mercuri and their technicians for providing the raw climate and NDVI data. Graciela Magrin and Ma. Isabel Travaso provided information on future climate scenarios for the region. Jimena Damonte helped with literature review and Andrea Goijman and Laura Solari provided useful comments on the manuscript.

REFERENCES

Barros, V., M. E. Castañeda, and M. Doyle. 1996. Variabilidad interanual de la precipitación: señales del ENSO y del gradiente meridional hemisférico de temperatura. In *Impacto de las Variaciones Climáticas en el Desarrollo Regional un Análisis Interdisciplinario*. VII Congreso Latinoamericano e Ibérico de Meteorología, pp. 321–322.

Benton, T. G., J. A. Vickery, and J. D. Wilson. 2003. Farmland Biodiversity: Is Habitat Heterogeneity the Key? *Trends in Ecology and Evolution,* 18(4):182–188.

Bibby C. J., N. D. Burgess, D. A. Hill, and S. Mustoe. 2000. *Bird Census Techniques.* London: Academic Press.

Blair, R. B. 1999. Land use and avian species diversity along an urban gradient. *Ecological Applications,* 6(2):506–519.

Brown, J. L., S-H. Li, and N. Bhagabati. 1999. Long-Term Trend toward Earlier Breeding in an American Bird: A Response to Global Warming? *Proceedings of the National Academy of Sciences,* 96:5565–5569.

Buckland, S. T., D. R. Anderson, K. P. Burnham, J. L. Laake, D. L. Borchers, and L. Thomas. 2001. *Introduction to Distance Sampling. Estimating Abundance of Biological Populations.* New York: Oxford University Press.

Canavelli, S., and M. E. Zaccagnini. 1996. Mortandad de Aguilucho Langostero (*Buteo swainsoni*) en la Región Pampeana: Primera Aproximación al Problema. INTA, Informe de Proyecto.

Canavelli, S. B., M. E. Zaccagnini, F. F. Rivera-Milan, and N. C. Calamari. 2003. "Bird Population Monitoring as a Component of Pesticide Risk Assessment in Argentine Agroecosystems." Proceedings of the 3rd International Wildlife Management Congress, 28 November–4 December 2003, Christchurch, New Zealand.

Calamari, N. C., S. B. Canavelli, and M. E. Zaccagnini. 2003–2007. Mapas de Abundancia Relativa de Aves en Agroecosistemas. www.inta.gov.ar/ and http://www.probiodivinta. com.ar/ templates/2column_c_f.asp?NIVEL=NIVEL3-A&opcion=2.

Caviglia, O. P., V. O. Sadras, and F. H. Andrade. 2004. Intensification of Agriculture in the South-Eastern Pampas: I. Capture and Efficiency in the Use of Water and Radiation in Double Cropped Wheat-Soybean. *Field Crops Research,* 87:117–129.

Clergeau, Ph. 1995. Importance of Multiple Scale Analysis for Understanding Distribution and for Management of an Agricultural Bird Pest. *Landscape and Urban Planning,* 31:281–289.

Dale, V. H. 1997. The Relationship between Land-Use Change and Climate Change. *Ecological Applications,* 7(3):753–769.

Donald, P. F., R. E. Green, and M. F. Heath. 2001. Agricultural Intensification and the Collapse of Europe's Farmland Bird Populations. *Proceedings of the Royal Society London, Series B,* 268:25–29.

[ESRI] Environmental Systems Research Institute. 1999. *User Guide ArcView GIS,* Version 3.2. Redlands, CA: Environmental Systems Research Institute, Inc.

———. 2005. ArcGIS 9.1. Redlands, CA.

Filloy, J., and M. I. Bellocq. 2007. Patterns of Bird Abundance along the Agricultural Gradient of the Pampean Region. *Agriculture, Ecosystems and Environment,* 120:291–298.

Fuller, R. J., R. D. Gregory, D. W. Gibbons, J. H. Marchant, J. D. Wilson, S. R. Baillie, and N. Carter. 1995. Population Declines and Range Contractions among Lowland Farmland Birds in Britain. *Conservation Biology,* 9:1425–1441.

Goldstein, M., B. Woodbridge, M. E. Zaccagnini, and S. B. Canavelli. 1996. An Assessment of Mortality of Swainson's Hawks on Wintering Grounds in Argentina. *Journal of Raptor Research,* 30:106–107.

Goldstein, M., E. T. E. Lacher, M. E. Zaccagnini, M. I. Parker, and M. Hooper. 1999a. Monitoring and Assessment of Swainson's Hawks in Argentina Following Restrictions on Monocrotophos Use, 1996–1997. *Ecotoxicology,* 8(3):215–224.

Goldstein, M., T. E. Lacher, B. Woodbridge, M. Bechard, S. B. Canavelli, M. E. Zaccagnini G. Cobb, E. J. Scollon, R. Tribolet, and M. Hooper. 1999b. Monocrotophos-Induced Mass Mortality of Swainson's Hawks in Argentina, 1995–1996. *Ecotoxicology,* 8(3):201–214.

Gordo. O., L. Brotons, X. Ferrer, and P. Comas. 2005. Do Changes in Climate Patterns in Wintering Areas Affect the Timing of the Spring Arrival of Trans-Saharan Migrant Birds? *Global Change Biology,* 11:12–21.

Gordo, O., and J. J. Sanz. 2006. Climate Change and Bird Phenology: A Long-Term Study in the Iberian Peninsula. *Global Change Biology,* 12:1993–2004.

Grau, H. R., T. M. Aide, and N. I. Gasparri. 2005. Globalization and Soybean Expansion into Semi-arid Ecosystems of Argentina. *Ambio,* 34(3):265–266.

Gregory, R. D., D. Noble, R. Field, J. Marchant, M. Raven, and D. W. Gibbons. 2003. Using Birds as Indicators of Biodiversity. *Ornis Hungarica,* 12/13:11–24.

Harrington, R., I. Wolwod, and T. Sparks. 1999. Climate Change and Trophic Interactions. *TREE,* 14(4):146–150.

Hoffman, J. A. J., S. E. Nuñez, and W. Vargas. 1997. Temperature, Humidity and Precipitation Variations in Argentina and the Adjacent Sub-Antarctic Region during the Present Century. *Meteorologische Zeitschrift,* 6:3–11.

Hooper, M., P. Mineau, M. E. Zaccagnini, and B. Woodbridge. 2002. "Pesticides and International Migratory Bird Conservation." In *Handbook of Ecotoxicology,* ed. D. J. Hoffman, B. A. Rattnes, G. A. Burton, and J. Cairns, pp. 737–753. Boca Raton, FL: CRC Press.

Hurtado, R., I. Barnatán , C. Mesina, A. Beltrán, and L. Spescha. 1996. Corrimiento de las Isoyetas Trimestrales Medias en la Región Pampeana Argentina, 1940–1990. Revista del IV Congreso Colombiano de Meteorología, pp. 141–146.

Hutto, R. L., and J. S. Young. 2002. Regional Landbird Monitoring: Perspectives from the Northern Rocky Mountains. *Wildlife Society Bulletin,* 30:738–750.

Lovett, G. M., D. A. Burns, C. T. Driscoll, J. C. Jenkins, M. J. Mitchell, L. Rustad, J. B. Shanley, G. E. Likens, and R. Haeuber. 2007. Who Needs Environmental Monitoring? *Frontiers in Ecology and Environment,* 5(5):253–260.

Magrin, G. O., M. I. Travasso and G. R. Rodriguez. 2005. Changes in Climate and Crop Production during the 20th Century in Argentina. *Climatic Change* 72:229–249.

Mineau, P. 2002. Estimating the Probability of Bird Mortality from Pesticide Sprays on the Field Study Record. *Environmental Toxicology and Chemistry,* 21(7):1497–1506.

Mineau, P. 2003. "Avian species." In *Encyclopedia of Agrochemicals,* ed. J. R. Plimmer, D. W. Gammon, and N. N. Ragsdale, pp. 129–156. Hoboken, NJ: Wiley Interscience.

Murphy, M. 2003. Avian Population Trends within the Evolving Agricultural Landscape of Eastern and Central United States. *Auk,* 120(1):20–34.

Naumann, M., and M. Madariaga. 2003. Atlas Argentino/Argentinienatlas, Programa de Acción Nacional de Lucha contra la Desertificación. Secretaría de Ambiente y Desarrollo Sustentable. Instituto Nacional de Tecnología Agropecuaria, Deutsche Gesellschaft für Technische Zusammenarbeit, 94 pp. Buenos Aires.

Nichols, J. D., and B. K. Williams. 2006. Monitoring for Conservation. *TRENDS in Ecology and Evolution,* 21(12):668–673.

Noss, R. F. 1990. Indicators for Monitoring Biodiversity: A Hierarchical Approach. *Conservation Biology,* 4(4):355–364.

Ojima, D. S., K. A. Galvin, and B. L. Turner. (1994). The Global Impact of Land-Use Change. *BioScience,* 44(5):200–204.

O'Connors, R. J., and M. Shrubb. 1986. *Farming and Birds.* Cambridge, UK: Cambridge University Press.

Panigatti, J. L., D. Buschiazzo, and H. Marelli. 2001. Siembra Directa II. Ediciones INTA.

República Argentina. 2007. 2da Comunicación Nacional de la República Argentina a la Convención Marco de las Naciones Unidas sobre Cambio Climático. Buenos Ares, Argentina

Root, T. 1988. Environmental Factors Associated with Avian Distributional Boundaries. *Journal of Biogeography*, 15:489–505.

Schrag, A. M., S. B. Canavelli, N. C. Calamari and M. E. Zaccagnini. 2009. Influence of Lands Use, Climate and Vegetation on Bird Species Richness and Composition in Central Argentina. *Agriculture, Ecosystem and Environment* 132:135–142.

Siriwardena, G. M., S. R. Baillie, S. T. Buckland, R. M. Fewster, J. H. Marchant, and J. D. Wilson. 1998. Trends in the Abundance of Farmland Birds: A Quantitative Comparison of Smoothed Common Birds Census Indices. *Journal of Applied Ecology,* 35:24–43.

Siriwardena, G. M, H. W. P. Crick, S. R. Baillie, and J. D. Wilson. 2000. Agricultural Land-Use and the Spatial Distribution of Granivorous Lowland Farmland Birds. *Ecography,* 23(6):702–719.

Solbrig, O. T. 1999. "Observaciones sobre biodiversidad y desarrollo agrícola." In *Biodiversidad y Uso de la Tierra*, ed. S. D. Matteucci, O. T. Solbrig, J. Morello, and G. Halffter, pp. 29–40. Buenos Aires, Argentina: Eudeba.

Stephens, P. A., R. P. Freckleton, A. R. Watkinson, and W. J. Sutherland. 2003. Predicting the Response of Farmland Bird Populations to Changing Food Supplies. *Journal of Applied Ecology,* 40:970–983.

Stevenson, I. R., and D. M. Bryant. 2000. Climate Change and Constraints on Breeding. *Nature,* 406:366–367.

Szaro, R. C., and B. K. Williams. 2008. "Climate Change: Environmental Effects and Management Adaptations." In *Climate Change and Biodiversity in the Americas*, ed. A. D. Fenech, D. MacIver, and F. Dallmeier, pp. 277–294. Toronto, ON: Environment Canada.

Thomas, L., J. L. Laake, S. Strindberg, F. F. C. Marques, S. T. Buckland, D. L. Borchers, D. R. Anderson, K. P. Burnham, S. L. Hedley, J. H. Pollard, J. R. B. Bishop, and T. A. Marques. 2006. *Distance 5.0. Release 2*. Research Unit for Wildlife Population Assessment, University of St. Andrews, UK. http://www.ruwpa.st-and.ac.uk/distance/.

Thompson, J. J. 2007. Agriculture and Gamebirds: A Global Synthesis with and Emphasis on the Tinamiformes in Argentina. Ph.D. diss., University of Georgia, Athens.

Viglizzo, E. F. 2001. La Trampa de Malthus—Agricultura, competitividad y medio ambiente en el siglo XXI, Editorial Universidad de Buenos Aires, pp. 90–91.

———. 2007. Desafíos y oportunidades de la expansión agrícola en Argentina. In.Pag. 12–41, Martínez Ortíz, U. Ed. Producción Agropecuaria y Medio Ambiente. Fundación Vida Silvestre Argentina.

Wiens, J. A. 1989. Spatial Scaling in Ecology. *Functional Ecology,* 3:385–397.

Wiens, J. A., B. Van Horne, and B. R. Noon. 2002. "Integrating Landscape Structure and Scale into Natural Resource Management." In *Integrating Landscape Ecology into Natural Resource Management*, ed. J. Liu and W. W. Taylor, pp. 23–67. Cambridge, UK: Cambridge University Press.

Winkler, D. W., P. O. Dunn, and C. E. McCulloch. 2002. Predicting the Effects of Climate Change on Avian Life-History Traits. *Proceedings of the National Academy of Sciences,* 99(21):13595–14599.

Woodbridge, B., K. K. Finley, and S. T. Seager. 1995. An Investigation of the Swainson's Hawk in Argentina. *Journal of Raptor Research*, 29:202–204.

Zaccagnini, M. E. 2006 ¿Porqué monitoreo ecotoxicológico de diversidad de aves en sistemas productivos? In *INTA Expone 2004, Conferencias presentadas en el Auditorio Ing. Agr. Guillermo Covas. Volumen III*, ed. J. Larrea, pp. 69–89. Buenos Areas, Argentina: Ediciones INTA.

Zaccagnini, M. E., and N. Calamari. 2001. "Labranzas Conservacionistas, Siembra Directa y Biodiversidad." In *Siembra Directa II*, ed. J. L. Panigatti, D. Buschiazzo, and H. Marelli, pp. 29–68. Buenos Areas, Argentina: Ediciones INTA.

Zak, M. R., M. Cabildo, and J. G. Hodgson. 2004. Do Subtropical Seasonal Forests in the Gran Chaco, Argentina, Have a Future? *Biological Conservation,* 7:35–44.

Impacts of Climate Extremes on Biodiversity in the Americas

Marianne B. Karsh[1,2] and Don MacIver[2]

ABSTRACT: Since 1970, climate extremes have been impacting biodiversity in the Americas with greater frequency, duration and severity than ever previously recorded. There is widespread evidence of longer droughts, more frequent wildfires, higher temperatures and more intense storms, hurricanes and precipitation events. As well as greater variability of El Niño Southern Oscillation events, the total area impacted by flooding, glacier retreat and permafrost melt, desertification, landslides and avalanches has grown. Further, concurrent extreme events—such as flooding and high temperatures, droughts and high winds, and droughts and flooding—are becoming increasingly common. Extreme events are not only emerging as a critical factor in climate change, but their correlation to predicted changes in biodiversity is greater than climate change alone. Although ecosystems show high resilience to hurricanes, ice storms and other extreme events, significant impacts on biodiversity may occur once certain thresholds in duration, intensity and severity are exceeded. The resulting losses in biodiversity can reduce ecological resilience and adaptive capacity to climate change. To manage for potential biodiversity loss and to provide adaptation options for ecosystems to become more resilient to climate hazards, researchers and policymakers require a baseline monitoring database. The current database, the forest biodiversity observing network, consists of more than 500 individual observing sites that allow for transect studies to interlink climate and biodiversity information. Canadian case studies are featured to illustrate the benefits of using transect studies to analyze the impacts of

[1]*Arborvitae, 804-60 Pleasant Blvd., Toronto, Ontario, Canada M4T 1K4.*
[2]*Adaptation and Impacts Research Division, Environment Canada, 4905 Dufferin Street, Toronto, Ontario, Canada M3H 5T4.*
Corresponding author: M. Karsh (Marianne.Karsh@ec.gc.ca).

climate hazards and their associated risks for biodiversity. To strengthen the existing database of biodiversity observing sites in the Americas, future sites located in areas of critical biodiversity, across climate, chemical and physical gradients, are discussed. Strategies for risk assessment and the analysis of impacts on biodiversity are shown for sites with hurricane, ice storm and heavy browsing damage.

Keywords: *biodiversity, climate extremes, hazards, warming, hurricane, ice storm, wildfire, drought, flood, diversity curves, synergies, ecosystems, conservation, resilience, adaptation, Smithsonian Institution, global monitoring network*

INTRODUCTION

The warming of the climate system is unequivocal, as is now evident from observations of increases in global air and ocean temperatures, widespread melting of snow and ice (glacial and permafrost), and rising global average sea level (IPCC, 2007; Loik et al., this volume). Within this new reality of climate change, the intensification of climate extremes has emerged as one of the most complex and critical factors, profoundly impacting species, ecosystems and the Earth's biodiversity as a whole (IPCC, 2001; Jentsch et al., 2007).

Small changes in climate can result in disproportionately great changes in frequency and magnitude of extreme events (Leemans and van Vliet, 2005). A shift in the distribution of temperature has a much greater effect at the extremes than near the mean. A shift of 1 standard deviation causes a 1-in-40-years event to become a 1-in-6-years event, reducing wait times between extreme events (Kharin et al., 2007). Thus, impacts will be more rapid, diverse and widespread than those previously predicted (Leemans and van Vliet, 2005).

Ecosystems respond faster to extreme weather than to average climate. There is a more rapid appearance of ecological responses to extreme events and most observed changes in biodiversity are attributed to changes in extremes (Leemans and van Vliet, 2005). Seasonal extent and climate variability are more of a factor in explaining species gradients than annual climatic conditions, with the exception of annual potential evapotranspiration (Githaiga-Mwicigi et al., 2002). The rate of climate change and the impact and pattern of climate extremes will be critical, and these will vary at national, regional and even local levels.

Ecosystems display amazing recovery responses to most climate extremes. In some cases, biodiversity may actually increase because of a reduction in the dominant species that allows for a greater number of nondominant species to occupy the site (Huston, personal communication, 2007). There may also be an increase in biodiversity as more invasive species and warmth-demanding plant species occupy sites impacted by climate extremes under a warming climate. But while a warmer climate can, in an ideal setup, eventually support more biodiversity (new species), it also leads to losses of native species.

Biodiversity will also decrease in the immediate to short-term when subjected to climate extremes, especially with decreased wait times between events. This is especially true of intense fires, hurricanes (category 3 or higher), heat waves, extreme precipitation and storm surges, prolonged drought and desertification. The biodiversity of species adapted to cold and wet conditions, species in high latitude areas, species with low reproductive rates and/or limited mobility, endemic species, and specialist species with narrow habitat requirements and long lifetimes is expected to decrease with increased climate extremes (Archaux and Wolters, 2006; Leemans and van Vliet, 2005, Table 1).

The threat of climate extremes to biodiversity is further compounded by the fact that humans have altered the structure of many of the world's ecosystems through habitat fragmentation, land degradation, pollution and other disturbances, making them less resilient to change (Szaro and Williams, 2008). The situation is critical, with biodiversity decreasing 1,000 times faster now than at rates found in fossil records (Balvanera et al., 2006). In the past, one mammal and two bird species were lost every 400 years; at present-day rates, 58 mammals and 115 birds are being lost in an equivalent time span (Groombridge, 1992). Species of mammals and birds reported to be extinct since 1970 include the Omiteme Cottontail, Toolache Wallaby, Dobson's Fruit Bat, Colombian Grebe, Bush Wren and the Guam Flycatcher (http://www.unep-wcmc.org/latenews/extinct.html).

Biodiversity and climate have a complex, symbiotic relationship. Not only does climate impact biodiversity but biodiversity, in turn, also mitigates climate extremes. Biodiversity is fundamental to climate regulation and is an important consideration in stabilizing ecosystems and supporting sustainable development (Stott, 2007). Ecosystems containing many different plant species serve as more efficient carbon sinks, are more productive and can better withstand and recover from climate extremes, pests and disease. As extreme events continue to escalate in intensity, the capacity of natural ecosystems to act as buffers to climate change, including climate extremes, is undermined. This has serious implications for ecosystems and societies (IPCC, 2001; Jentsch et al., 2007).

How ecosystems cope with escalating climate extremes is thus a central question in climate change and ecological conservation (Reush et al., 2005). The International Panel on Climate Change reports a measurable increase in intense tropical cyclone activity and areas impacted by drought since 1970 (IPCC, 2007). The same trend is evident for wildfire and severe floods. Most current climate change studies indicate that the frequency, duration, severity and intensity of extreme events will continue to increase and that there will be more climatic variability (Conway, 2007; Markham, 1996; Jentsch et al., 2007; Szaro and Williams, 2008; Table 1). With such intensive changes impacting ecosystems, it is imperative to understand these changes and to track responses to climate extremes across a range of ecosystems and bio-climatic zones in the Americas.

A comprehensive review of the scientific literature was compiled to identify current and anticipated impacts of six major categories of climate extremes on biodiversity in the Americas: extreme heating, hurricanes, drought, ice storms, wildfire, and precipitation and floods. Analysis of these findings, as well as data obtained from a standardized assessment of diversity in existing one-hectare biodiversity sites, reveals how dramatically climate extremes are reshaping ecosystems and reinforces the need for a global monitoring organization to track changes and make appropriate adaptation recommendations.

MEASURING THE IMPACT OF CLIMATE EXTREMES

Gauging the impact of climate extremes on biodiversity requires reliable quantitative forms of measurement. Diversity curves are a rapid assessment tool that allows scientists to consider all species and not focus solely on the rare ones. They can be used to illustrate climate-related

TABLE 1. Impacts of climate extremes on biodiversity in the Americas.

Climate Extremes	Prediction	Impact Areas (Americas)	Impacts on Biodiversity	Source
Extreme weather events	• Increasing in frequency, duration and severity • More common in the last 30 years • Increases as change accelerates with continued warming • Smaller diurnal and seasonal changes	• North, South and Central America • Unique biomes, small island states, irreplaceable features/wet tropics, border ranges (limited places to shift), land masses below cloud base in Caribbean	• Increasing adverse risk to biodiversity • Strong correlation with changes in biodiversity accounts for most of observed changes • More rapid appearance of ecological response • Extremes will intensify already existing trends • Initial benefits followed by adverse impacts • Species not able to adapt or migrate in some cases if climatic thresholds or tolerances exceeded • Not able to sustain high levels of human impact for long periods • Increased demand for water; water stress with increased temperatures	Markam, 1996 Dallmeier & Comiskey, 1998 IPCC, 2001 Krockenberger et al., 2003 Leemans & van Vliet 2005 Killeen, 2007
El Niño Southern Oscillation events	• Increasing in variability • Current models do not predict increase; others suggest an increase in magnitude and frequency	• Midwest USA, Central America and Pacific Coast, Peru, Ecuador, Bolivia, Mexico, Caribbean, Brazil, Colombia, Peru, north Amazonia	• Indirect cause of landslides, floods, droughts, hurricanes • May be trigger for ice storms in eastern Canada • Attributed to increased climate variability • Increased dry conditions in some regions • Increased risk of forest fires • May cause desertification in dryland ecosystems • Decline in grasslands with variable precipitation	Breshears et al., 2005

Climate Extremes	Prediction	Impact Areas (Americas)	Impacts on Biodiversity	Source
High air temperatures	• Increasing day and night extreme temperatures • Increasing frequency of extreme temperatures • Higher maximum and minimum temperatures • More hot days • Reduced diurnal temperature ranges	• All across North, Central and South America: Alaska, Banks Island, Alberta, Washington, Colorado, California, Olympic Mountains, Montana, Southeast Arizona, Mexico, Central America, Peru, Argentine Islands, Ecuador, Bolivia, northern Chile, Andes Mountains, Colombia • Raised peat bogs, alpine/mountains, Arctic, Boreal forest, arid and semiarid areas	• Increased Net Primary Productivity (NPP) in northern forests; decreased NPP in eastern USA and tropics • Productivity may increase with small rise in global mean temperature; shows peaks and declines • Change in areas of vegetation types • Decline or death of tropical or temperate forests (high temperature) • Species extinctions (high temperature) • Water stress at higher CO_2 concentrations • Initial benefit followed by adverse impacts • May reduce cold resistance of plants/trees and negatively influence species responses to high temperatures • Southern ecological boundaries more impacted than northern • Gypsy moth and pear thrips more aggressive; malaria, dengue fever increasing • Earlier budding, leafing or blooming; earlier egg-laying, spawning, emergence from hibernation • Polewards or upwards shift of treeline; animal range shifts	Markam, 1996 IPCC, 2001 Krockenberger et al., 2003 Leemans & van Vliet, 2005 Malcolm et al., 2006 Fischlin, 2007 Jentsch et al., 2007 Killeen, 2007 Varrin et al., 2007

(Continued)

TABLE 1. Impacts of climate extremes on biodiversity in the Americas. (Continued)

Climate Extremes	Prediction	Impact Areas (Americas)	Impacts on Biodiversity	Source
High air temperatures			• Disruption in animal migrations; altered competitive balances • Warmth-demanding species more abundant in last 30 years • Endemic species replaced by generalists (better competitors) • Small changes can have large impacts on biodiversity • Coral reef bleaching (warmer water temperatures) • Range contractions and shifts • Longer growing season, accelerated maturation rates • Effects on sex determination in many species (incubation temperatures) • New species found in northern latitudes, high mountainous regions • New unusual weather patterns in northern regions: thunder; lightning • Genetic adaptation of insects	Markham, 1996 Allen & Breashears, 1998 Harrison, 2000 IPCC, 2001 Krockenberger et al., 2003 Peters et al., 2004 Breshears et al., 2005 Archaux & Wolters, 2006
Drought	• Increasing in extent, severity, frequency and duration • Total area impacted by drought expanding • Increased summer drying • Increased desertification and aridity	• Eastern and Southwest USA: New Jersey, Delaware, Maryland, Rhode Island, West Virginia, Florida, Texas, Louisiana • Central America and Caribbean • South American coastline	• Drying of river beds • Increased risk of some pests, pathogens, disease (e.g., bark beetles) • Mortality (prolonged extreme drought—over 15 months) • Increased heat stress	

Climate Extremes	Prediction	Impact Areas (Americas)	Impacts on Biodiversity	Source
Drought		• Northeastern South America • Southern Peru • Southwest Bolivia • Northeast Brazil • Chile, Argentina • Northern Mexico • Tropical montane forests	• Increased risk of forest fires and water shortages; accelerates soil erosion • Premature leaf fall, dieback, wilting of trees, deterioration of vegetative cover • Some taxa increase; less adapted taxa decrease • Good resistance to drought—but duration of drought a factor • May increase biodiversity by reducing dominance of most abundant plants • Increased risk of deforestation • Increased risk of increased use of chemicals • Altered phenology, reduced flowering and reproductive success • Reduced below and above ground biomass • Reduced cover of perennials; rapid colonization of annuals and biennials • Increased rates of species extinctions • Increased risk of windthrow • Decreased crop yields and food security • Increased risk of dust storms when combined with high winds • Greatest impact on very young or old plants, species adapted to cold or wet conditions, and species with low reproductive rates or limited mobility	Huston, personal communiation 2007 Jentsch et al., 2007

(Continued)

TABLE 1. Impacts of climate extremes on biodiversity in the Americas. (Continued)

Climate Extremes	Prediction	Impact Areas (Americas)	Impacts on Biodiversity	Source
Desertification	• Increasing; expansion of existing areas	• Arid and semiarid regions	• Reduced biological productivity • Change of vegetation area: encroachment of woody plants into perennial grasslands • Land degradation	Peters et al., 2004
Fire	• Increasing in frequency, severity, duration, size, amplitude • Increase in areas burned • Increase in areas with severe fires • Earlier start to fire season and longer fire seasons • Shorter fire return intervals	• Both USA and Canada have seen increases in areas burned in the last 35 years • West Central and northwestern Canada in boreal forest and taiga, • St. Lawrence region • Western USA, Florida • Western coast of South America: Argentina • Mexico, Nicaragua (Bosawas BR)	• Possibly most harmful on biodiversity: more rapid rate of vegetation transformation than other climate-induced change • Seedling establishment hindered • Plant and animal habitat modified • Mortality of fire-sensitive species (rainforests, vine forests). • Changes distribution of dominant plant species; affects understory • Bares mineral soil; younger age distributions and earlier successional stages may result • Converts temperate dry forests to grasslands and moist temperate forests to dry woodlands; transforms vegetation (lower NPP) • Increases risk of invasive species and feral animals • Impacts streams and fisheries; higher stream temperatures • Highest impact in forest plantations, shrub and open forest areas; degraded forests, temperate forests, boreal forest, taiga	Eversham & Brokaw, 1996 Markham, 1996 Dallmeier & Comiskey, 1998 Stocks et al., 1998 Krockenberger et al., 2003 Odion et al., 2004 Lavorel et al., 2007 Betts, 2007 Killeen, 2007

Climate Extremes	Prediction	Impact Areas (Americas)	Impacts on Biodiversity	Source
Heat waves	• Increasing in frequency, intensity, duration and severity • Increase in heat index • Increase in dryness	• USA: Rockies, North and South Dakota, Illinois, New York, Maine, Arkansas, Texas	• Increased heat stress and mortality (month-long heat waves) • Increased extinction of sensitive species, mortality as temperatures rise above tolerable thresholds for too long a period • Moose not able to withstand high temperatures	IPCC, 2001 Krockenberger et al., 2003
Extreme precipitation	• Increasing frequency of extreme showers • Increasing variability and intensity; more wet periods • Decreasing in subtropics • Increasing in northern regions	• USA: New England, California, Texas • Venezuela, Uruguay, Argentina, Brazil, Paraguay, Bolivia, Peru, Haiti • Tropical montane forests (cloud cover), raised peat bogs	• Increased difficulty of ecosystems dealing with increased variability of precipitation events • Increased biomass production by plants in semiarid USA • May result in increased deer and mice populations and increased browsing stress • May change competitive ability among plants • May be related to earlier fire season	IPCC, 2001 Krockenberger et al., 2003
Storms	• Increasing in frequency, duration and intensity • Storms more widespread	• Central America, Caribbean, Pacific Coast: linked to warmer, wetter climate and less sea ice (open water provides a source for moisture, heat) • Low lying islands, boreal forests, coral reefs, mangroves, marshes	• Damage to coral reefs • Habitat damage • Increased frequency and intensity of storms shifts ecosystems to less desired stage, with diminished capacities to generate ecosystem services followed by loss of ecosystem resilience	Markam, 1996 IPCC, 2001 Folke et al., 2004 Archaux & Wolters, 2006

(Continued)

TABLE 1. Impacts of climate extremes on biodiversity in the Americas. (Continued)

Climate Extremes	Prediction	Impact Areas (Americas)	Impacts on Biodiversity	Source
Floods + high temperatures	• Increasing	• Rivers, wetlands	• Decline in aquatic and riparian diversity • Decrease water in rivers and wetlands • Increase frequency of water shortages • Negative impact on river and wetland ecosystems	Krockenberger et al., 2003
Dryness + high temperatures	• Increasing in certain regions	• Tropical regions	• Dieback of humid tropical forest: plants absorb less carbon through photosynthesis than is released through soil respiration, accelerating further warming • Livestock and wildlife prone to heat stress; rainforests suffer	Krockenberger et al., 2003
Drought + high winds	• Increasing	• Tropical regions	• Increased dust storms; loss of soil and its biodiversity	Krockenberger et al., 2003
Hurricanes, catastrophic winds, cyclones	• Increasing (more category 4 and 5) • Possible reduction • More frequent, intense and widespread • Triggered by El Niño-type weather patterns and ocean warming • Increase in tropical cyclone peak wind and precipitation intensities	• Caribbean Islands • Most of South America • Gulf of Mexico, North Atlantic • Tropical montane forests, coral reefs and mangroves particularly vulnerable to increasingly intense tropical cyclones	• One of the most important disturbances after fire: destroy forests, change competitive interactions among plants and alter successional pathways • Reduced diameter, basal area, number of stems, biomass • High adaptive response • Full recovery rate long (hundreds of years)	Markam, 1996 Eversham & Brokaw, 1996 Dallmeier & Comiskey, 1998 Arevalo et al., 2000 Krockenberger et al., 2003 Huston, personal communication, 2007 Jentsch et al., 2007 IPCC, 2007

Climate Extremes	Prediction	Impact Areas (Americas)	Impacts on Biodiversity	Source
Hurricanes, catastrophic winds, cyclones			• Increased biodiversity: less dominance of abundant species	
			• Significant sprouting; multi-stemmed trees	
			• Reduced above-ground biomass; increased NPP (disturbances release nutrients); complete defoliation	
			• Long-term impact on forest structure and function, less desirable storm forests, more impact on dry slow-growing forests	
			• Reduced wind resistance	
			• Prevents pioneers reaching canopy (early mortality for large stems) due to mortality, wind damage, falling debris	
			• Increased fire, landslides	
			• Diversity maintained (alternate mechanism for forest turnover) but possible decrease between events	
			• Increased range in gaps; increased light	
			• Accelerates succession; survival rates highest in species with rapid growth	
			• Increased damage in even-aged forests	
			• Increased freshwater runoff	
			• Loss of habitat	

(Continued)

TABLE 1. Impacts of climate extremes on biodiversity in the Americas. (Continued)

Climate Extremes	Prediction	Impact Areas (Americas)	Impacts on Biodiversity	Source
Floods	• Increasing in severity, frequency, intensity and duration • Increased variability of river flows and water availability • Increasing risk of winter floods and coastal flooding • Increasing number of flood disasters; may alternate with drought	• Canada, USA, Mexico • Caribbean: Haiti, Dominican Republic • Central America: Belize, El Salvador, Costa Rica • South America: north Peru, southwestern Bolivia, eastern Brazil, Chile, Argentina, Uruguay, Paraguay, Venezuela, Colombia, Guyana, Ecuador • Coastal habitats, beaches, mangrove areas, deforested areas on steep slopes	• Land inundation • High sedimentation • May stimulate lush vegetation • Deterioration of vegetative cover	IPCC, 2001 Krockenberger et al., 2003
Ice storms, heavy snow events, frost, cold waves, cold events	• Decreasing frequency and magnitude of extreme cold events • Milder winters, increasing in magnitude and extent • Sharp cooling; frosts increasing • Frequency and size of ice storms increasing as more heat is trapped in lower atmosphere	• Eastern Canada and USA: Canadian Arctic; Mount Baker, Washington; Black Hills, South Dakota • South America: Argentina, Brazil	• Mortality with 75% crown damage • High adaptivity to damage, high rate of recovery at lower levels; epicormic branching • Recently thinned conifer plantation more affected • Less flexible species (maples, oaks) more affected than more flexible species (pines, birches)	Neilsen et al., 2003 Krockenberger et al., 2003 McCready, 2004 Jentsch et al., 2007 Huston, personal communication, 2007 Varrin et al., 2007

Climate Extremes	Prediction	Impact Areas (Americas)	Impacts on Biodiversity	Source
Ice storms, heavy snow events, frost, cold waves, cold events			• May increase biodiversity by reducing dominance of most abundant species • Reduced access to forest because of flooding, snow, wind- and ice-damaged trees • Direct damage to trees by wind, snow, ice • More diverse stands are best form of insurance • Severity depends on ice accumulation, duration, area affected, wind speed; timing may be more important than temperature • Caribou deaths: snow and freezing rain bury food sources • May result in sudden cold injuries that are often lethal • May destroy forests, change competitive interactions among plants, alter successional pathways and advance succession	

(Continued)

TABLE 1. Impacts of climate extremes on biodiversity in the Americas. (Continued)

Climate Extremes	Prediction	Impact Areas (Americas)	Impacts on Biodiversity	Source
Ice and snowmelt, glacier retreat, permafrost melt	• Increasing (especially in the last 25 years) • Sea surface temperatures increasing • Fewer cold extremes • Milder winters • Increased winter rainfall	• Bering Sea, Arctic Ocean, Alaska, • Canada: Hudson Bay; Rockies • USA: New Hampshire; Glacier National Park, Montana • South America: Colombia, Ecuador, Peru, Bolivia, Venezuelan Andes, Argentina • Polar regions, high mountains	• Sharp drop in summer water flows; increasing erosion • Increased frequency of landslides and avalanches; increased risk of floods and mudslides • Local biodiversity adversely impacted; sea bird decline • Loss of ice sheets • Increased risk of sea level rise, negatively affecting mangroves, which buffer against flooding, high winds and erosion • Increased risk of winter river floods • Earlier river thaw, spring ice-out and spring snowmelt (may affect fire regimes) • Decreased sea ice and snow; increased thawing of permafrost • Glacier retreat • Reduced water availability with melting of tropical glaciers, leading to increased desertification and aridity, increased pests and diseases and increased impact on soybean production • Earlier spring discharge from snowmelt	Markam, 1996 IPCC, 2007 Fischlin, 2007 Conway, 2007

Climate Extremes	Prediction	Impact Areas (Americas)	Impacts on Biodiversity	Source
Sea level rise, coastal flooding, storm surges	• Increasing, becoming more rapid • Influenced by increasing ice melt and coastal storms • Collapse of West Antarctica ice sheet might impact sea level suddenly • Triggered by rising air and ocean temperatures	• USA: Hawaii, West Coast, California, Louisiana, Florida Keys, Chesapeake Bay • Caribbean: Bahamas, Bermuda • Mexico • Central America: Panama, Costa Rica, El Salvador, Belize • South America: Argentina, Brazil, Colombia, Galapagos, Guyana, Ecuador, Peru, Chile, Venezuela, Uruguay • Mangroves, coral reefs, coastal marshes, low-lying islands, deltas	• Saltwater intrusion of aquifers, estuaries • Beach habitat lost • Increased land inundation • Slow but steady loss of wetlands and drylands • Saltwater inundation of coastal mangrove forests: since 1980, 20% of mangroves gone; 25% more fish on reefs close to mangroves gone • Half of all agricultural lands may be subject to salinization & desertification by 2050 • Negative impact on fish stocks • Increased storm surges • Wildlife reserves submerged	CBD, 2006

(Continued)

TABLE 1. Impacts of climate extremes on biodiversity in the Americas. (Continued)

Climate Extremes	Prediction	Impact Areas (Americas)	Impacts on Biodiversity	Source
Ocean warming and ocean acidification	• Increasing	• USA: West Coast, California, Florida Keys • Caribbean: Bahamas, Bermuda • Mexico • South America: Panama, Galapagos, Ecuador	• Decline in zooplankton populations • Coral reef ecosystem collapse • Increased competition from algae • Coral reef bleaching (since 1980) • Increasing risk of hurricanes • Increasing risk of extreme events • Sea bird decline • Decreased dry season mist: may result in species extinction and declines of frogs, toads and lizards • Shoreline and species shifts northwards • Upward shift in mountain birds • Negative impact on any shell-producing species	Markham, 1996 IPCC, 2001
Landslides, avalanches, earthquakes volcanoes	• Increasing in frequency • Landslides influenced by increasing deforestation on steep slopes in South America	• USA • Mexico • Caribbean: Haiti, Puerto Rico, Dominican Republic • Central America: Guatemala, El Salvador, Honduras, Nicaragua, Costa Rica, Panama, Coastal Ecuador • South America: Venezuela, northern Peru, Guyana	• Reduced above-ground biomass; increased NPP (disturbances release nutrients) • Debris flow and mudslides • Global climate impacted	Krockenberger et al., 2003 Varrin et al., 2007

impacts, with the y-axis representing proportional abundance on a logarithmic scale and the x-axis representing the number of species or families. The families are ranked from greatest to least abundance, with each family having a minimum of five individuals.

Diversity curves provide a relative measure of proportional abundance of trees between sites, identifying areas of high, moderate and low abundance. Latitudinal gradients of diversity exist from high in the tropics to low in the Arctic. Peru averages 152 different plant species per hectare, with a diameter of 10 cm or more. In northern Europe, the average is 18 species per hectare, and in the eastern United States, it is 29. In Canadian sites, the average number of tree species is 11 per hectare. The diversity curves can also show if, and to what extent, the number of families might be reduced through climate extremes, as well as other threats such as human impact and prescribed burns (Environment Canada, 2003).

Image 5 (see photospread) shows diversity curves for tree families generated from biodiversity data for the hurricane-impacted site at Bisley, Puerto Rico (vis-à-vis comparable diversity at Backus Woods, Long Point Biosphere Reserve, Canada and Soberania National Park, Panama), highly diverse Urubamba, Peru, moderately diverse Jiangfengling National Park, China and Dikola, Camaroon, and single-species forests in Charlevoix Biosphere Reserve, Canada. Black spruce at Charlevoix is an important Canadian monoculture benchmark. However, this is only one tree family per hectare compared to nearly 50 in South America. Backus Woods, Canada and Bisley in Central America have a similar number of families (<15), while Dikola, Cameroon, has around 30 families per hectare (Image 5, see photospread). The diversity curves suggest that the conservation of a single species in a northern ecosystem may be more critical to the way that an ecosystem functions than it would in a highly diverse tropical ecosystem, with its abundance of species and genetic variations.

The diversity curves provide a benchmark for establishing existing biodiversity under current climate conditions. To see how climate extremes may affect this biodiversity, sites for which data are available have been evaluated to determine if they experienced a climate extreme during the monitoring period. The diversity curves were then compared before and after the event and also with similar sites that had not experienced a climate extreme. Since the existing network of 1-ha biodiversity monitoring plots were sites of opportunity, established in biodiversity reserves, parks or university lands and not necessarily in places where climate extremes were expected to occur, there is not enough data to generate diversity curves for the majority of the climate extremes examined here. The resulting gap in data reveals the importance of strategically locating sites in ecosystems vulnerable to climate extremes.

LITERATURE REVIEW AND ANALYSIS OF IMPACTS OF CLIMATE EXTREMES

Extreme Warming

Global air temperatures are expected to increase by 1.4 to 5.8°C by the end of the century, a rate of warming almost certainly without precedent in the last 10,000 years (IPCC, 2001). Increasing frequency and occurrence of higher day and night extreme temperatures are predicted. It is estimated that 3% of species will do well under the new conditions, 9% will experience no change, 15% will be able to adapt and 73% will not be able to adapt (Fischlin, 2007). Heat-stressed and degraded systems are expected to be replaced by better adapted ones but degradation is fast and recovery is slow. The recovery process is impacted by habitat fragmentation, pollution and other land-use changes. All regions of the Americas will be impacted, particularly raised peat bogs, the arctic, alpine and mountainous areas, the boreal forest, and arid and semiarid ecosystems.

Warmth-demanding plant species have become more abundant in the last 30 years, coinciding with a precipitous rise in temperatures. More generalist species that are better able to compete and opportunistic species with wide ranges and rapid dispersal are likely to become more abundant, while endemic species and specialists with narrow habitat and long lifetimes are in danger of declining. There will likely be an increase in weedy plants or invasive species because of their faster ability to reproduce. High temperatures may also reduce cold resistance of plants and negatively influence species responses to variability in temperature extremes. With extremely high temperatures, there is a risk of species extinctions and decline or death of tropical or temperate forests.

Canadian ecosystems and communities are already experiencing the challenge of adapting to a changing climate, with average temperatures in Canada's north warming at rates some two to three times greater than the rest of the world. Canada's northern ecosystems, indigenous cultures and population health are experiencing significant impacts from the changing climate and from other related atmospheric changes, including increasing persistent organic pollutants and ultraviolet radiation levels. Northern ecosystems will be significantly impacted by an expected 50% melting of the permafrost and decreased snowfall (Gough and Leung, 2002). Animals, including polar bears and some bird species, may be extirpated in Ontario, Canada, as their range shifts north (Varrin et al., 2007).

In tropical Central and South America, there is a high risk of bird loss and significant species extinctions in the near future in areas with the most significant concentration of biodiversity. With a 1°C increase in global air temperature, 50% of large primates and 9% of tree species are at risk of extinction and the function and composition for forests is modified (IPCC, 2001; MEA, 2005). A 1°C change can reduce boreal forests by 25% and cause major dieback in these and other forests (Markham, 1996). With forest dieback in the Amazon, there is less recycling of rainfall and increased CO_2 in the atmosphere (Betts, 2007), as well as a massive increase in abundance of lianas because of CO_2 fertilization that is concomitant with rising temperatures. This increase in lianas is less efficient as a carbon sink as it results in smaller forest carbon stores (Betts, 2007).

Several biomes in Costa Rica with the highest numbers of endemic species are especially vulnerable to extremes in temperatures and may experience the largest reductions in area or will disappear completely (reviewed by Malcolm et al., 2006). El Niño Southern Oscillation has been identified as the cause of decreasing precipitation and increasing temperature in Costa Rica. When these climate extremes are combined with deforestation, a rising of the cloud cap will probably occur. In the cloud forests, ferns and orchids will die with the lifting of the cloud cap, and toad and frog populations could go extinct (Markham, 1996; Killeen, 2007). Loss of trees in cloud forests is irreversible. If trees are cut, no vegetation is able to gather moisture from clouds and conditions are too dry for the forest to reestablish (Wilson and Agnew, 1992; Folke et al., 2004). Preserving sufficient forest cover in the Amazon Basin to maintain rainfall patterns requires conservation efforts and monitoring across the entire ecoregion (Olson et al., 2001).

Extremes in temperature will exacerbate many of the impacts on biodiversity already observed. Ranges of flying insects, birds and raccoons have moved north, arctic and alpine plants have contracted northward, and genetic changes have been noted in insects, birds and mammals. Breeding dates have shifted for frogs, spring peepers and birds. Arctic ice is thinner, breaking up earlier and forming later. Other impacts that will be compounded by temperature extremes include increased local diebacks and extinction rates of trees, earlier budding and leafing, and a poleward or upward shift of the treeline border between trees and tundra (IPCC, 2001). The maximum rates of spread for some sedentary species, including large tree species, may be slower than the predicted rates of change in climatic conditions.

Studies have shown that the urban climate of the Toronto, Ontario, Canada has already warmed, equivalent to the expected 2050 climate. To illustrate the impact of extreme temperature on biodiversity, a 1-ha climate change experimental site was established in 2002 at the Humber Arboretum in Toronto.

Along with native tree and shrub species, trees native to Washington, D.C.—a region two growing zones to the south, the distance that trees might potentially have to migrate under a change in temperature extremes—were also planted at the site. On this site, tree survival and growth was mitigated by extreme browsing damage. Milder winters have resulted in an increase in deer and mouse populations. What was surprising was the extent of the browsing damage, despite protection from tree guards and other measures, with over 70% of the newly planted trees browsed and approximately 30% mortality in five years (M. B. Karsh, preliminary analysis, 2008). One of the lessons of this study is that temperature extremes in the urban forest need to be offset with the planting of much larger stock to reduce browsing impact from the concomitant increase in animal populations.

Hurricanes

The number of hurricanes, storms and catastrophic wind events are all expected to increase and to become more intense. A greater number of category 4 and 5 hurricanes are predicted, as these are linked to El Niño weather events and ocean warming. Since category 3 or higher hurricanes can result in a restructuring of the forest, Caribbean islands, tropical montane forests, coral reefs and mangroves are all at risk (Krockenberger et al., 2003).

The immediate and severe effects of hurricanes are complete defoliation and mortality. Most of the damage is a direct result of wind or falling debris, so wind protection in these forests is crucial. Hurricanes may prevent pioneer tree species from living long enough to reach the forest canopy and large individuals of early successional species have the highest damage rates (Dallmeier and Comiskey, 1998). This creates room for late successional hardwoods (Arevalo et al., 2000). Species with rapid growth following the hurricane are the best survivors.

The "storm" forests in the Caribbean are characteristic of the type of impact hurricanes can have: short, shrubby, multistemmed trees with spongy heartwood. A shrubby multistemmed forest structure seems to be a natural response to hurricane winds. There is an increase of ethylene production at the root collar resulting in more auxin being produced and more hormones, which results in a sprouting response (van Bloem et al., 2006). Sprouting maximizes leaf area index and minimizes the distance to transport water to the stem. These forests are typified by reduced number of canopy strata, abundance of lianas, absence of giant trees and dominance by one or two species (Dallmeier and Comiskey, 1998).

An increased frequency and intensity of storms and hurricanes may shift ecosystems to a less desired stage, with diminished capacities to generate ecosystem services, followed by loss of ecosystem resilience and increased risk of forest fire (Folke et al., 2004). Forests adapted to extreme events recover quickly, but it may be decades before the site will attain the floristic, physiognomic and avifaunal characteristics of the mature forest (Dallmeier and Comiskey, 1998).

Image 6 (see photospread) shows the Bisley site in Puerto Rico before and after Hurricane Hugo hit in 1989. There was a reduction in the number of families, from nine to eight, a year after the hurricane. By 1999 there were 12 or more families on this site, which is comparable to other sites in the region. It appears that the hurricane initially reduced biodiversity, but that recovery took place quite quickly.

Severe Drought

The regions impacted by drought are predicted to expand and summer drought is expected to increase in frequency, extent and duration before the end of this century (Archaux and Wolters, 2006; Fischlin, 2007; Loik et al., this volume). Tropical montane forests will be particularly affected. These forests support ecosystems of distinctive floristic and structural form and contain a disproportionately high number of the world's endemic and threatened species. Drought in some regions of the Americas may be accompanied by increasing aridity and desertification (Markham, 1996). As summer soil moisture decreases, drought stress increases.

Maximum moisture deficits are expected in northwestern Argentina, the southwestern United States and northwestern Mexico. Central California and the intermountain west may have significant deficits. North Chile and central and southeastern Argentina are predicted to be dry and the area affected by very dry conditions may double as a result of climate change and climate extremes (Laurenroth et al., 2004). The evergreens in the Amazon could be transformed to savanna ecosystems within the next century (Killeen, 2007). Savannization of Amazonia and drier biomes in northeast Brazil, along with accompanying desertification, is anticipated because of the synergistic combination of climate change, climate extremes and land-use changes (Nobre, 2007).

Forest biodiversity may be particularly resistant to drought, although the duration of drought may be the most important factor when assessing impact on ecosystems. The relationship between the timing of drought, flowering and pollinators is critical. A delay in flowering due to drought can lead to extinction of pollinators (Harrison, 2000). Prolonged drought can also cause desertification, enabling greater encroachment by woody plants into perennial grasslands and associated land degradation in arid and semiarid regions (Peters et al., 2004). Further impacts of desertification include reduced biological productivity and chronic wildfires.

Extreme drought acts synergistically with other threats. Infestation by bark beetles is closely linked with drought-induced water stress (Allen and Breshears, 1998; Breshears et al., 2005). With increased temperatures, host defenses are lowered, allowing pathogens and parasitism to flourish. Extremes can negatively precondition trees and thereby increase their susceptibility to secondary damage through pests and pathogens and make them more susceptible to windthrow. Drought combined with high winds leads to increased dust storms and loss of soil and its biodiversity (Krockenberger et al., 2003).

Ice Storms

Most climate models project a decrease in frequency and magnitude of extreme cold events. There is also some evidence of sharp cooling and the frequency and size of ice storms may increase in some latitudes (Varrin et al., 2007). Minimum temperature effects can result in sudden cold injuries that are often lethal; however, the timing of extreme frost events may be more important than temperature (Jentsch et al., 2007).

Ice storms result in reduced access for people in the sugar maple industry to forestland because of flooding, deep snow, or wind- and ice-damaged trees. Mortality of trees is indicated at 75% crown damage. Trees have a high rate of recovery at lower damage levels and changes in composition and loss of syrup production are nonpersistent (McCready, 2004). Recently thinned forest plantations, single species ecosystems, and maples and oaks are generally the most impacted by ice; however, susceptible species are often balanced by epicormic branching (rapid crown recovery). As with other extremes, more diverse stands are the best form of insurance against impacts (Neilson et al., 2003).

In January 1998, an area from Kingston, Ontario, to the Maritimes in Canada was struck by the worst glaze ice storm of the century. The impact of the storm was concentrated in the valley of the St. Lawrence River, with the zone around St.-Jean-sur-Richelieu, southwest of Montreal, Quebec, being the most heavily hit. Prior to the ice storm, four permanent forest biodiversity monitoring sites had been established in the affected region: two sites at the Mont St. Hilaire Biosphere Reserve, Quebec; a site at Gananoque, Ontario; and a site at Ste-Hippolyte, Quebec. After the storm, a quantitative assessment of the damage to all four 1-ha sites was conducted. The Mont St. Hilaire Biosphere Reserve lay within the zone of greatest damage. The range of crown damage for this storm was 20%–50% (Environment Canada, 2003).

Images 7a and 7b (see photospread) show diversity curves for ice storm–damaged pure sugar maple sites in comparison to undisturbed sugar maple and also to more biologically diverse sites. The maple mixed wood sites at Gananoque, Tiffin, and Long Point, Ontario, Canada, have very similar patterns of proportional abundance, with the two Carolinian sites at Long Point having 9 to 11 tree families and the maple mixed wood sites at Tiffin and Gananoque with 6 and 8 tree families, respectively (Image 7a, see photospread).

The two sugar maple sites in Quebec, whether damaged in the ice storm or not, had four tree families (Image 7b, see photospread). The ice storm damaged forest at Gananoque, Ontario, with eight families was younger and different in composition than the two impacted sites at Mont St. Hilaire, Quebec, with almost half the number of families. The maple mixed wood site in Mont St. Hilaire had five tree families in comparison to the maple mixed wood site in Gresse-lle Historic Site, Quebec, with seven tree families. It is unclear whether this reduction in families could be attributed to the ice storm since less than 70% of the crown was damaged.

Wildfire

Forest fire is associated with global climate change and climate extremes (Lavorel et al., 2007). As the maximum number of consecutive dry days is decreasing in the Caribbean and elsewhere, the number of areas with severe wildfires is expected to rise in the Americas by the end of this century. Fire seasons are expected to lengthen by as much as 30 days in some regions facilitated by increased spring and winter temperature extremes and earlier spring snowmelt (Miller, 2006). There is an anticipated earlier start to the fire season, with a significantly greater number of areas experiencing severe to extreme fires.

Already a trend of increasing fire size, amplitude and duration has been observed (Odion et al., 2004). The area burned by forest fires in Canada has risen dramatically since 1970 as summer season temperatures have warmed. The areas burned in the United States have also been on the rise since 1960 (Fischlin, 2007). Large and severe wildfires are predicted to become more common. The areas particularly at risk are boreal forest, taiga, temperate forests, managed stands and forest plantations, streams and fisheries.

Extreme forest fire modifies biodiversity and can result in significant changes in distribution of dominant plant species and younger age distributions (Stocks et al., 1998). Extreme fire also affects the understory and bares mineral soil (Lavorel et al., 2007). In this respect, its impact exceeds those of other extreme events (Eversham and Brokaw, 1996), and may incite a more rapid rate of vegetation transformation than other climate-induced change. It can cause temperate dry forests to become grasslands and moist tropical forests to become dry woodlands. With extreme fire, seedling establishment is hindered, plant and animal habitat is modified and species are at greater risk.

Fire is synergistically linked to El Niño events, hurricanes, drought, lightning strikes, deforestation, forest fragmentation and land-use change. Extreme forest fire is most severe during El Niño years, when drought conditions predominate (Killeen, 2007). Fires in the tropics often follow hurricanes. Since these forests are often not adapted to fire, more than 90% of all canopy trees can be destroyed and the understory completely burned. Even so, recovery may occur with vegetation repopulating the site in five years time but with different species, causing the forest to be in an earlier successional stage and therefore less able to withstand climate extremes in the future. Fire is more prevalent in degraded forests, so any extreme or human impact that reduces the resiliency of the forest puts it at higher risk of extreme fire. Large forest fires release tons of carbon dioxide into the atmosphere, which acts as an escalating trigger for even more climate extremes and fire.

Intense Precipitation and Floods

The frequency and variability of extreme precipitation showers is increasing. This puts extreme stress on ecosystems that can more readily adjust to reduced precipitation than greater variability in precipitation. Areas vulnerable to extreme and variable precipitation are tropical montane forests, peat bogs, Arctic, arid and semiarid regions. Intense showers have particularly high impact in places like Venezuela, where the risk of erosion may be higher on steep deforested slopes.

The number of flood disasters is expected to increase and these may alternate in the same regions as droughts. Storms and floods affect low-lying islands, boreal forest, coral reefs, mangroves, coastal marshes, fisheries, animal habitat, sandy beaches, and ecosystem resistance. The combination of floods and high temperature events is on the rise, especially impacting rivers and wetlands. Ocean warming and ocean acidification is also escalating, affecting coral reefs, zooplankton populations and shell-producing organisms. Glacier retreat and permafrost melt is becoming more extensive and is affecting polar, mountain and alpine regions. This type of extreme is also linked to greater risk of landslides, avalanches, floods, mudslides, sea level rise and reduced water availability. Sea level rise and storm surges are increasing impacting aquifers, estuaries, fisheries, agriculture, and wildlife reserves through salt water intrusion, land inundation and habitat destruction.

SYNERGIES BETWEEN CLIMATE EXTREMES AND OTHER THREATS TO BIODIVERSITY

The impacts of climate extremes will depend on other significant processes, such as habitat loss and fragmentation. The top threats to biodiversity that are exacerbated by climate extremes include global climate change, habitat destruction, deforestation, invasive species, and fire regime alteration (Ervin and Parrish, 2006; Killeen, 2007). For cerrado vegetation in Brazil, high rates of habitat destruction to cropland suggest that only current reserves will survive (Thomas et al., 2004) and the cerrado may disappear completely by 2030 since it is very suitable for mechanical agriculture (Killeen, 2007).

According to a literature review and assessment by the World Wildlife Fund (WWF) (2007), summarized in Tables 2 and 3, the primary threat to biodiversity is deforestation in Brazil, the Caribbean, Chile, Colombia, El Salvador, Guyana, French Guyana, Mexico, Panama, Peru and Venezuela. This implies that any international effort on introducing incentives for avoided deforestation in Central and South America may have huge benefits.

Human impacts decrease ecosystem resiliency and increase vulnerability to climatic hazards (Betts, 2007). Discrete events of novel extreme magnitude and frequency may drive ecosystems beyond stability and resilience (Jentsch et al., 2007). Resiliency may be exceeded this century because of climate change, climate extremes, land use changes, pollution and overexploitation of resources (high confidence IPCC4—Fischlin, 2007). The rapid rate of climate change, climate extremes and a fragmented landscape will inhibit the ability of many species to adapt or migrate to regions adequate for survival (Killeen, 2007). Key climate variables may increase in frequency to a point at which species are unable to recover normal or viable populations (Markam, 1996). All of the above listed threats are challenges for mitigation and adaptation and may involve reassessment of human economic activity. A description of a few of the major impacts and why they are of concern, along with recommended adaptations, are listed in Table 4.

BASELINE MONITORING DATABASE

The synergistic interactions between climate extremes and other hazards are impacting ecosystems to such an extent that it is imperative for a global monitoring network to track these changes and assess ecosystem resiliency. To manage for potential biodiversity loss and to provide adaptation options to help ecosystems become more resilient to climate hazards, researchers and policymakers require a baseline monitoring database. Organizations such as

TABLE 2A. Hazards to Biodiversity by Country (South and Central America).

Hazards to Biodiversity	Argentina	Bolivia	Brazil	Caribbean	Chile	Columbia	Costa Rica	Cuba	Dominican Republic	Ecuador
Agriculture	4	5	15	14	2	7	3	6	2	4
Agri-industry (palm oil, charcoal, peat, salt ext.)			3	1		2		1		1
Animal/plant extinctions			. 5	3	1	1	2			1
Avalanches						1				
Biodiversity loss	1		1	2	1	1		2		
Buildings and infrastructure	1			4		4			1	
Climate warming—air, ocean temp.			1							1
Deforestation/logging	3	3	20	16	6	10	3	6	2	3
Desertification	*		1		1					
Disease						1				1
Displacement of indigenous people										
Drought			1	*						
Earthquake				*	*		*		*	*
Fires—wildfire and human set fires/ slash and burn	3	2	10	7	2	1	1	4	1	1
Habitat destruction and fragmentation	1	3	5	6	1	4	1	1		
Hunting and poaching	6	3	4	5	1	4	1	2	2	4
Hurricanes				*						
Illegal wildlife and plant trade	3	4	2	2		1		4	1	1
Introduced plants/animals and invasive species	1	2	4	10	5	2	1	2		2
Land conversion	2	2	6	6	2	2	1	1		

Hazards to Biodiversity	Argentina	Bolivia	Brazil	Caribbean	Chile	Columbia	Costa Rica	Cuba	Dominican Republic	Ecuador
Lightening			1			1				
Mudslides						1				
Mining	2	1	14	7	2	1	1	3		1
Narcotic crops (opium, heroin, marijuana)					2	1				
Oil extractions/gas/oil spills	2		2	2		2				1
Overexploitation of plants and animals	1	1	7	1	2	1		5		3
Overfishing	1		3	3		3		1		1
Population increases and settlements		3	7	8	1	5	2			2
Public awareness/conflicts			1	9						
Ranching and grazing	7	1	18	6	4	6	3	4	2	6
Roads	4	3	7	6	3	3				3
Sea level rise				3				1		
Soil erosion	1		3	6	2	3	5	1		
Tourism	2		7	7	2			2	1	2
Tropical storms				*					*	
Urbanization/development	2		13	9	2	5	3	3		1
Volcanic eruptions				*		2				
War						1				
Waste—improper dumping	2		1	1	1	1	1	1	1	1
Water draining/salinization/diversions		2	2			2	1	2	1	1
Water pollution	1	1	8	2	2	6		1	1	1
Water shortages/sedimentation	1		3	2		2	3	3	1	1

(Continued)

TABLE 2A. Hazards to Biodiversity by Country (South and Central America). (Continued)

Hazards to Biodiversity	Argentina	Bolivia	Brazil	Caribbean	Chile	Columbia	Costa Rica	Cuba	Dominican Republic	Ecuador
Number of Reported Sites in Literature	11	5	42	21	8	16	6	6	2	6
Disasters Report Scale										
Desertification	4									
Earthquake				5	5		5.5		5	4.5
Drought				4						
Hurricanes				5						
Lightening										
Tropical storms/floods				4.5					4	
Volcanic eruption				4						

* Very significant occurrence; see disasters report scale.

Caribbean countries not mentioned separately in this table include Anguilla, Antigua, Bahamas, Barbados, Cayman Islands, Dominica, Haiti, Martinique, St. Vincent, Trinidad.

TABLE 2B. Hazards to Biodiversity by Country (South and Central America).* (Continued)

Hazards to Biodiversity	El Salvador	French Guiana	Guatemala	Guyana	Honduras	Mexico	Nicaragua	Panama	Peru	Venezuela
Agriculture	1	1	1		1	17	2	3	7	8
Agri-industry (palm oil, charcoal, peat, salt ext.)						3				
Animal/plant extinctions					1	9		1	2	1
Avalanches										
Biodiversity loss						7				3
Buildings and infrastructure						5			1	3
Climate warming—air, ocean temp.				1		1		1		
Deforestation/logging	2	1	1	2	1	22	1	5	7	8
Desertification						1				
Disease										
Displacement of indigenous people										1
Drought										1
Earthquake	*		*		*			*		
Fires—wildfire and human set fires/slash and burn				1	1	6			1	4
Habitat destruction and fragmentation		1	1	1	1	11		1	4	4
Hunting and poaching		1		1	1	13	1		3	6
Hurricanes							1			
Illegal wildlife and plant trade				1		10		1	1	4
Introduced plants/animals and invasive species						4		1		2
Land conversion		1				8		3	1	3
Lightning	*									
Mudslides										
Mining		1		2	1	4	1	1	6	4
Narcotic crops (opium, heroin, marijuana)								1		
Oil extractions/gas/oil spills		1			2	2		1		6

(Continued)

TABLE 2B. Hazards to Biodiversity by Country (South and Central America). (Continued)

Hazards to Biodiversity	El Salvador	French Guiana	Guatemala	Guyana	Hondu-ras	Mexico	Nicara-gua	Panama	Peru	Venezuela
Overexploitation of plants and animals						12		3	3	5
Overfishing						4		1	2	1
Population increases and settlements	2				1	9	1	3	1	5
Public awareness/conflicts						6				
Ranching and grazing	1	1				18	1	3	4	6
Roads						10		3	1	6
Sea level rise	1				1	9	1	3	2	1
Soil erosion	1	1		1		7	1			6
Tourism				1			1			
Tropical storms							1			
Urbanization/development	1	1	2		1	12	1	4	3	5
Volcanic eruptions										
War	1			1	1	2	1		1	
Waste—improper dumping	1				1	7			1	
Water draining/salinization/diversions	1			1	1	9		1	3	6
Water pollution	1	1				9	2	1	3	4
Water shortages/sedimentation	1	1			2	5	1		2	1
Number of Reported Sites in Literature	2	1	2	2	1	36	2	5	10	13
Disasters Report Scale										
Desertification										
Earthquake	5.5	4.5		4			4.5			
Drought										
Hurricanes										
Lightening	4									
Tropical storms/floods										
Volcanic eruption										

¹Caribbean countries not mentioned separately in this table include Anguilla, Antigua, Bahamas, Barbados, Cayman Islands, Dominica, Haiti, Martinique, St. Vincent, Trinidad.

TABLE 3. Summary of threats to biodiversity by country (South and Central America).

Country	Primary Concerns	Secondary Concerns	Geo-hazard
Argentina	Ranching, grazing, overhunting	Agriculture, road construction, desertification	
Bolivia	Agriculture, illegal wild-life trade	Logging, road construction, population increases, hunting, habitat destruction	
Brazil	Deforestation, ranching	Mining, urban encroachment and development	Species are showing impacts from increased temperatures, drought and desertification
Caribbean	Deforestation, agriculture	Ranching, fires, urbanization, introduced plants and animals	Earthquakes, volcanic eruptions, tropical storms and drought
Chile	Deforestation, introduced plants and animals	Ranching	Earthquakes
Colombia	Deforestation	Ranching, agriculture, water pollution	This is an area that is vulnerable to avalanches and mudslides, frequently triggered by soils that have been destabilized as a result of logging
Costa Rica	Soil erosion	Deforestation, sedimentation, ranching, urbanization, agriculture	Severe earthquake zone
Ecuador	Ranching, agriculture, habitat destruction	Deforestation, overexploitation of plants and animals	Fairly severe earthquake zone
El Salvador	Deforestation, population pressure		Fairly severe earthquake zone; also subject to lightning strikes

(Continued)

TABLE 3. Summary of threats to biodiversity by country (South and Central America). (Continued)

Country	Primary Concerns	Secondary Concerns	Geo-hazard
Guatemala	Urban encroachment		Earthquakes
Guyana and French Guiana	Deforestation, mining	Overhunting, water pollution, habitat destruction, tourism	
Honduras and Nicaragua	Agriculture, water pollution	Deforestation, soil erosion, sedimentation, urbanization, hunting, improper dumping of waste	Earthquakes
Mexico	Deforestation, ranching, agriculture	Urbanization, hunting, overexploitation of plants and animals, habitat destruction	
Panama	Deforestation, urbanization	Soil erosion, population increases, ranching, road construction, agriculture, land conversion, overexploitation of plants and animals	
Peru	Deforestation, agriculture	Mining	
Venezuela	Deforestation, agriculture	Water diversions, ranching, roads, hunting, oil extraction, tourism	

TABLE 4. Climate change impacts and potential adaptations for biodiversity.

Hazards and Climate Change Impacts	Biodiversity Applied Adaptation
Agriculture: Agriculture is the largest driver of land-use change in the Americas. Already, 50% of the Cerrado ecosystem has been converted to farmland. Highly mechanized modern agricultural techniques are able to overcome limitations of tropical soils for agriculture, enabling agriculture to keep expanding wherever there is arable land (Killeen, 2007).	An increase in the biodiversity of the farming system may help to reduce the contribution to drought. Banks of drought- resistant, flood-resistant, saline-resistant seed varieties can be used to respond to climate extremes. Protect pollinators and their habitat: 35% of the world's crops depend on pollinators such as bees.
Aquaculture: Mangroves are being converted to shrimp farms at very high rates; 35% of mangroves have already been lost (MEA, 2005).	Protect and replant mangrove areas.
Administrative problems and lack of financial resources	Prioritize protection of coastal regions, pollinators, endemic species, rainforests and wetlands. Provide legal protection for lands. Regulate lands and commodity markets to reduce or eliminate complete deforestation of the Amazon. Implement efficient regulatory measures to counter market forces (Killeen, 2007). Reward agricultural producers for conservation of wetlands and for provision of ecological goods and services. Reward forest conservation by providing incentives. Offer financial compensation for areas of avoided deforestation: payment for the environmental service of not harming the rainforest. Pay rent for carbon storage.
Biofuels: Biofuels are among the greatest threats to the conservation of Amazon wilderness areas and hotspots. The projected market for biofuels is so large that it could stimulate deforestation far beyond the most pessimistic scenarios (Killeen, 2007). Sugar cane is grown to produce alcohol and oil palm to produce biodiesel oil. The African oil palm and elephant grass, which are ideally suited for use as a biofuel flourish and replace natural vegetation in the tropics. Oil palm emits isoprene, which produces more ozone, resulting in detrimental effects (Betts, 2007).	Assess the difficult tradeoff with biofuels. Biomass from diverse prairies can make biofuels without annual tilling, fertilization and pesticides. Perennials have more root mass than crops such as corn, which must be replanted annually are more stable, have fewer requirements and can still produce biofuels.
Decrease in extent and thickness of sea ice	Reduce greenhouse gas emissions.

(Continued)

TABLE 4. Climate change impacts and potential adaptations for biodiversity. (Continued)

Hazards and Climate Change Impacts	Biodiversity Applied Adaptation
Dams: Flooding of reservoirs causes declines in fish diversity (Killeen, 2007).	Avoid mega-hydroelectrical projects on primary rivers (Killeen, 2007).
Deforestation: Deforestation is occurring at very high rates across the Americas, with Brazil having the highest absolute forest loss. The loss of butterfly habitat in Mexico and the loss of gallery forests in Amazonia are of particular concern. A savannah ecosystem requires cover for wildlife and is too exposed if the adjacent gallery forests disappear (Killeen, 2007). Forests in Central and South America are often converted for other uses such as agriculture, ranching and biofuels. Deforestation in these regions is occurring in a climate where precipitation is decreasing and temperature is increasing compounding the stress on trees. Deforestation affects the process of evapotranspiration, which reduces precipitation in the ecosystem and expands arid conditions. All of these conditions exacerbate tree mortality. Deforestation also entails an increase in emissions of greenhouse gases. This is one of the greatest challenges to biodiversity facing us (Stott, 2007).	Restore peatlands: peat bogs sequester twice as much CO_2 as all of the world's forests. Implement avoided-deforestation programs. Lower the annual deforestation rate by 5%: this will reduce disasters and the avoided emissions will result in a $650-million-a-year credit (Killeen, 2007). Forests aid in the amelioration of climate extremes. Plant vegetation to protect against floods and coastal erosion. Plant mangrove belts and tree shelter belts (Conway, 2007; IPCC, 2001). Establish timber harvest cycles of 100 years to allow ecosystems to recover between extreme events. Practice proactive planting: the hundreds of years that it takes for natural biodiversification to occur is too long given the impacts and stresses coming with anticipated rates of climate change and extremes (Environment Canada, 2003).
Desertification	Plant species tolerant to higher temperatures.
Drought	Increase variety of species; include succulents in cropping system. Plant drought-resistant seed varieties. Keep river corridors forested as this makes ecosystems more resilient to drought.
Extinctions of plant and animal species	Practice in-situ and ex-situ conservation. Maintain genetic as well as species diversity to enhance ecosystem resilience (Reusch et al., 2005). The death of species is the responsibility of all. We have to act urgently.
Eutrophication/Decreased water quality	Plant/protect species important to improving water quality, e.g., mangroves.
Flooding	Plant mangroves to diffuse tropical storm waves.

Hazards and Climate Change Impacts	Biodiversity Applied Adaptation
Fragmentation: Fragmented landscapes may not allow for the redistribution of populations to accommodate shifting climates. Edge effects penetrate at least 300 m, so remnants smaller than this in area generally result in biodiversity loss. Forest fragmentation and land degradation increases risk of wildfire. Fire and logging that follow in the wake of fragmentation allow for regeneration of pioneer species and an increase in invasive species. Isolation and fragmentation can also lead to extinction of endemic species and homogenization of biota in protected areas (Killeen, 2007).	Establish conservation corridors designed to promote biodiversity conservation. Conserve the hills, ridges and valleys in Brazil and Guyana as wildlife reserves.
Hunting: Intensive hunting is causing local extinctions of vertebrate fauna in the Americas. As seed dispersers are reduced, tree reproduction is negatively affected (Killeen, 2007).	Work with local communities to reduce hunting to sustainable levels. Work with timber companies to reduce logging associated hunting and wildlife trade. Work with governments to provide scientific data on hunting impacts.
Greenhouse gas emissions	Need rapid implementation of technology and strategies for carbon sequestration to decrease greenhouse gas emissions (Thomas et al., 2004).
Increases in extreme events: storms/fires	Proactively plant more resistant species according to anticipated impact. Diverse ecosystems cushion against climate uncertainty, are more productive, better withstand and recover from climate extremes, pests and disease, and increase stability.
Increased permafrost melt	Reduce greenhouse gas emissions.
Increased pressures on habitats	Maintain natural diversity of species; reduce fragmentation and human impact.
Infectious diseases	Reduce encroachment on animals' natural habitats. Establish preventative measures in areas where humans and wildlife live in close proximity.
Inundation of fresh water	Plant salt-tolerant species.
Invasive species	Control or eradicate invasive species.

(Continued)

TABLE 4. Climate change impacts and potential adaptations for biodiversity. (Continued)

Hazards and Climate Change Impacts	Biodiversity Applied Adaptation
Loss of migratory wildfowl and mammal breeding and forage habitat	Create new reserves with flexible boundaries. Conserve large blocks of forest to provide for the minimum distance for edge effects and to give rare species adequate range area for survival. Special consideration should be given to the West Amazon—the most diverse and most stable climate. Use corridors to connect ecosystems.
Lack of respect for ecosystems: Human lack of respect and care for ecosystems is reflected in the quality of water resources.	Education is very important, as is the participation of all stakeholders. Need to see environmental rights as human rights. Human beings have an ethical duty to conserve Earth's species, which have an intrinsic value over and above the benefits they provide. Conservation should be seen as a moral obligation to preserve our priceless, irreplaceable heritage (Killeen, 2007). Value species for their aesthetic, spiritual and intellectual inspiration.
Mining: Diamond, gold, copper and coal are all mined in the Americas; mining is even allowed in national parks in Bolivia. Diamond mining in the Northwest Territories affects migration routes for caribou. Mining interrupts the high and low water flows in rivers, reduces sediment loads and disrupts the migratory behavior of fish (Killeen, 2007). It also pollutes and destroys vegetation and wildlife, leaving residues that impact neurological function and increase risk of birth defects.	Implement land-use planning. Develop alternative technology. Restore natural sites where there has been mining. Enact regulations to protect reserves and parks.
Oil and gas extractions/oil spills: Oil spills are particularly problematic in Bolivia and Peru, where high rainfall and unstable topography increases their risk of occurring (Killeen, 2007).	Pursue alternative energy sources. Use improved production technologies. Phase out older oil tankers.
Overgrazing	Encourage sustainable grazing management techniques.
Reduced colonization success	Assist species regeneration.
Reduced water supply	Plant drought-tolerant species.

(Continued)

Hazards and Climate Change Impacts	Biodiversity Applied Adaptation
Reduction in wetland areas	Preserve wetlands because they control floods by storing large volumes of water and help reduce erosion by slowing runoff. They also have the potential to remove and store greenhouse gases from the Earth's atmosphere. Wetlands are important for waterfowl and key spawning, nursery and feeding areas for fish. They provide habitats for 600 species of wildlife.
Roads: Highway construction in the cloud forests of the Amazonia leads to extinction of species in areas of high endemism. Improved highways lead to deforestation, affecting both carbon emissions and continental precipitation patterns (Killeen, 2007).	Implement land-use planning to reduce impact.
Urban encroachment and development/population expansions: With urban expansion, vast wetlands are threatened by human encroachment.	Implement land-use planning to reduce impact. Educate the public and raise awareness.
Water depletions: Water depletions cause rivers to run dry, groundwater tables and aquifers to drop, and lakes to vanish. Freshwater depletions are occurring in the Everglades. Lake level declines and a reduction in wetlands are also taking place in Mexico. The deterioration of wetlands, rivers and riparian habitats has a disproportionately high impact on biodiversity. The loss of more than two-thirds of natural river flow results in a loss of biodiversity. This has a serious effect on the long-term health of the ecosystem, including on fish and other organisms (Krockenberger et al., 2003).	Enact regulations. Plant trees and introduce water conservation measures.

the Smithsonian Institution are recognizing the need for such a monitoring network and are starting to put a framework in place.

In 1992 the Smithsonian Institution (Dallmeier, 1992) initiated a global biodiversity monitoring program under the auspices of UNESCO. In the Americas, this forest biodiversity observing network consists of nearly 500 individual observing sites in 20 countries (Image 8, see photospread). The design of the sites recognizes that biodiversity conservation is not single-species management but, rather, the ability to manage multi-taxa simultaneously. It is also able to support the four major types of habitat biodiversity monitoring activities: monitoring based on species at risk; monitoring based on population trends; monitoring based on status and trends in habitat; and monitoring based on threats to biodiversity. By incorporating standardized 1-ha plot sizes, measurement protocols for multi-taxa and transects across physical, chemical and ecological gradients, the forest biodiversity observing network is able to facilitate unique investigations into the cumulative impacts of climate extremes on forest biodiversity that can increase understanding of climate change and help reduce the adaptive deficit of the Americas (Fenech et al., 2005).

Among the important findings that these sites can help yield are baseline data on climate variability and extremes in ecosystems. The heat-unit-by-family-biodiversity model for forests, suggested by Rochefort and Woodward (1992), is a useful means for obtaining such data.

Heat is a powerful trigger of climate extremes. Changes of 1° or 2°C average annual global temperature translate into significant biological impacts, adaptations and vulnerabilities. The heat unit is the accumulation of heat above the base relationship temperature of 5°C commonly referred to as a growing degree day (GDD). This basic relationship can serve as an effective diagnostic tool for gauging expected versus observed climate variability, and helping to identify sites where biodiversity is or is not in equilibrium with the changing climate (MacIver, 1998). The spatial variability of climate-based biodiversity is mapped and then subjected to a $2 \times CO_2$ atmosphere, using climate change scenarios to calibrate and again map the anticipated and future changes in biodiversity (MacIver, 1998).

Image 9 (see photospread) shows the number of species and families as related to heat units in Canada compared to eight sites in Virginia, USA, and sites in Panama, Ecuador, Peru, Bolivia, Paraguay, Argentina, Brazil and Venezuela. The South American sites help to provide the upper limit for forest biodiversity at the family levels along with heat units calculated for the Americas (Karsh et al., 2007). Based on sites in Canada, the USA, and Central and South America, preliminary research seems to suggest that the Family Diversity Axis may not be linear, but either logarithmic or exponential approaching equatorial regions (MacIver, 1998; Karsh et al., 2007).

In Canada, the heat-unit-by-family-biodiversity model has not only enabled initial verification of the family biodiversity baseline but also allowed for the calibration of biodiversity for mixed wood forest species in Southern Ontario. In addition, it has helped to identify areas of Southern Ontario that will require enhanced conservation practices for the adaptability of native species, including areas that will be vulnerable to climate extremes and other concurrent impacts such as invasive exotic species.

Image 10 (see photospread) depicts the gradient analysis and the relative linearity of the basic heat by family biodiversity relationship for mixed wood forest sites from Long Point on Lake Erie to Tiffin on Georgian Bay, Ontario, Canada (MacIver, 1998).

POTENTIAL NEW GLOBAL BIODIVERSITY MONITORING SITES ACROSS THE AMERICAS

The most vulnerable ecosystems to climate extremes will include those habitats where the first or initial impacts are likely to occur and those where the most serious adverse effects may arise or where the least adaptive capacity exists. These include Arctic, mountain and island ecosystems. Tools and guidance in the form of scientific predictions of ecological states are needed to highlight priority ecosystems at high risk of climate extremes and to guide climate change adaptation options.

To meet the full range of monitoring requirements, permanent plot networks must be expanded across the Americas to include sites subject to a broader range of climate extremes, human activities and impacts. Protected areas (IUCN), Global Biosphere Reserves (UNESCO) and Smithsonian global reference sites, based on scientifically sound, standardized monitoring protocols, are essential to provide an effective, community-based platform to monitor changes in ecosystem functioning and resilience In Ontario, Canada, analysis showed the tremendous value of using transect studies in evaluating climate impacts on biodiversity. Moreover, with only 11 families at the Canadian network's most diverse location in Long Point, Ontario, it further underscored the fact that Canada can ill afford to lose one species. In Ontario and Quebec, Canada, and in Bisley, Puerto Rico, the analysis has also highlighted the value of using Smithsonian sites established prior to the climate extreme event, allowing for direct comparisons of impacts.

It is recommended that more forest biodiversity observing sites be established in accordance to the Smithsonian Institution protocol across ecological, chemical and climate gradients in highly altered landscapes and vulnerable areas in the Americas to detect change and adaptation response to climate extremes. With sites from the equatorial to polar, climate-biological transects using standardized monitoring and global modeling could be used to further calibrate this analysis. The forest biodiversity global site networks, established to evaluate climate extremes under increasing human impact, would increase their coverage by being connected using GIS, remote-sensing-type technology.

The WWF has ranked ecoregions in the Americas by distinct biodiversity features: species endemism, rarity of higher taxa, species richness, unusual ecological or evolutionary phenomena, and global rarity of habitat type (WWF, 2007). This is particularly helpful when identifying areas to monitor. Areas in urgent need of protecting are regions in which monitoring sites need to be established. The areas identified as being in need of protection are lowland and montane forests in western Amazonia (Thomas et al., 2004), as well as the Pantanal, Gran Chaco, and Caatinaga regions (Killeen, 2007). The best remaining refugia for lowland moist forest species is on the western edge of Amazonia, adjacent to the Andean highlands. It is vital to strengthen and expand protected area and monitoring networks in the western Amazon. As much forest should be conserved in Amazonia as possible, since species ranges are expected to change drastically (Miles et al., 2004).

Some areas and species are highlighted for protection, monitoring and restoration simply for their overall significance and contribution. Peat bogs sequester twice as much CO_2 as all the world's forests even though they form only 3% of the global land area. Willow absorbs eight ton of CO_2 in its first growing season. Thirty-five percent of the world's crops depend on pollinators such as bees. Tropical forests have 50% of all plant and animal species, two-thirds of known terrestrial species and the largest number of threatened species. Vertebrates in these forests are very sensitive to abundance of food reserves, birds and mammals are affected by unusual weather, and the flowering of food plants and survival of trees is impacted by drought. Thus, all of these ecosystems need to be monitored very closely as their total impact on ecosystem resilience is high.

THE CASE FOR A UN GLOBAL MONITORING ORGANIZATION

Climate extremes will have a powerful impact on biodiversity and on the human communities that benefit from this biodiversity. Because of the strong linkages between biodiversity and climate extremes, there are many instances at regional, national and international levels where integrated joint actions that address the synergies between the issues are more effective than dealing with each issue separately.

It is expected that biodiversity's natural adaptation responses to changing climate, especially given other environmental pressures such as habitat loss and fragmentation, will be insufficient to stem the additional losses of biodiversity expected because of climate change. As a result, planned or directed adaptation activities are urgently needed now to slow the increased rates of future biodiversity losses. As an adaptation strategy, maintaining biodiversity allows ecosystems to provide goods and services to communities while societies learn to cope with climate change. Adaptation in support of biodiversity conservation is also essential if United Nations Framework Convention on Climate Change (UNFCCC) objectives and Millennium Development Goals for poverty alleviation, food production and sustainable development are to be met.

The integration of adaptation and mitigation actions within the context of climate extremes, biodiversity conservation and sustainable development all call for greater synergy in implementing the Convention on Biological Diversity (CBD) and other relevant multilateral environmental agreements. Synergies in protecting biodiversity and protecting the climate through adaptation measures will be facilitated by proposed joint action on biodiversity and mitigation measures through international pilot measures to reduce greenhouse gas emissions from deforestation and adaptation to deal with the changing climate impacts.

Planned adaptation encompasses efforts to restore resilience to ecosystems, since resilient ecosystems maintain biodiversity, continue to deliver ecosystem goods and services and protect human, plant and animal communities from climate hazards such as erosion, flooding and drought. It is recognized that a wide range of adaptation activities have been designed or planned by many countries including Asia, Africa, Kribati, Sudan, Pacific Islands, Nepal, Bangladesh, United States, Canada, Mexico, Columbia, United Kingdom and Finland (CBD, 2006), but that few such adaptation actions have been implemented to date. In many cases, adaptation actions have been delayed because the knowledge base and partnership processes required to support adaptation implementation for biodiversity needs to be strengthened. If adaptation- and resilience-building actions are delayed, studies indicate that many additional species will be lost, while management options will become more limited and expensive and have a lower likelihood of success.

Given that climate extremes and land degradation are major causes of biodiversity loss and that biodiversity conservation and its sustainable use can contribute to both climate-change mitigation and adaptation, it is necessary to have an organization responsible for promoting synergy at local, national and international levels. Desertification and land degradation issues affect global climate change through soil and vegetation losses, while biodiversity in turn influences carbon sequestration and therefore helps to regulate climate change. Since climate extremes pose one of the most significant threats to biodiversity, activities that promote adaptation to climate change can also contribute to the conservation and sustainable use of biodiversity, sustainable land use and water management.

All nations in the Americas must be encouraged to recognize the importance of developing synergistic actions on biodiversity, adaptation and climate change that maximize opportunities for successful biodiversity conservation and that also help to slow the rate of loss of biodiversity. Maintaining and enhancing biodiversity should be part of all national policies, programs and plans for adaptation to climate extremes in order to allow ecosystems to continue providing necessary goods and services, including support for natural disaster preparedness. Such a policy would increase investment in climate-change adaptation actions that will reduce the loss of biodiversity and strengthen resilience in natural ecosystems. One area that shows promise is the issue of potential incentives for the conservation of forests (avoided deforestation) and peatlands. Activities that slow deforestation and/or forest degradation could provide substantial biodiversity benefits and help to meet the global goal of reducing the rate of loss of biodiversity by 2010. The creation of a new organization at the UN level, responsible for consolidating and increasing the Global Biodiversity Monitoring Network and integrating biodiversity and climate information, would help immeasurably in developing and facilitating such urgently needed pan-American responses to climate change and extremes.

The disaster that struck New Orleans after Hurricane Katrina in 2005 illustrates that not only better dykes but healthy and resilient ecosystems are our best defense against the impacts of climate extremes. Hurricane Katrina touched down on a coastal region of the United States that has been under environmental pressure for over a century. Reengineering of the Mississippi River, accomplished through a system of canals and levees, diverted natural sedimentation flows and steadily eroded coastal wetlands. Development also destroyed barrier islands and oyster reefs that buffered the coast. During the hurricane, the tidal surge was able to travel unimpeded up shipping canals and burst over the levees surrounding New Orleans. Although damage from the storm would have been considerable in any case, breaches occurred more often in areas where wetlands had been destroyed and levees were exposed to wave action. A global monitoring organization is essential to establishing when and where such disasters may happen and to determine when ecosystems are out of equilibrium so that action can be taken before the critical tipping point.

REFERENCES

Allen, C. D., and D. D. Breshears. 1998. Drought-Induced Shift of a Forest-Woodland Ecotone: Rapid Landscape Response to a Climate Variation. *Proceedings of the National Academy of Sciences*, 95:14839–14842.

Archaux, F., and V. Wolters. 2006. Impact of Summer Drought on Forest Biodiversity: What Do We Know? *Annals of Forest Science*, 63:645–652.

Arevalo, J. R., J. K. DeCoster, S. D. McAlister, and M. W. Palmer. 2000. Changes in Two Minnesota Forests during 14 Years Following Catastrophic Windthrow. *Journal of Vegetation Science*, 11:833–840.

Balvanera, P., A. B. Pfisterer, N. Buchmann, J. He, T. Nakashizuka, D. Raffaelli, and B. Schmid. 2006. Quantifying the evidence for Biodiversity Effects on Ecosystem Functioning and Services. *Ecology Letters*, 9:1146–1156.

Betts, R. 2007. Biodiversity—Ecosystem and Climate Functioning. Presentation at Biodiversity-Climate Interactions, Royal Society, London, 12 June 2007.

Breshears, D. D, N. S. Cobb, P. M. Rich, K. P. Price, C. D. Allen, R. G. Balice, W. H. Romme, J. H. Kastens, M. F. Floyd, J. Belnap, J. J. Aderson, O. B. Myers, and C. W. Meyer. 2005. Regional Vegetation Die-Off in Response to Global-Change-Type Drought. *Proceedings of the National Academy of Sciences*, 102(42):15144–15148.

[CBD] Convention on Biological Diversity. 2006. Guidance for Promoting Synergy among Activities Addressing Biological Diversity, Desertification, Land Degradation and Climate Change. Ad hoc Technical Expert Group on Biodiversity and Adaptation to Climate Change, CBD Technical Series No. 25.

Conway, G. 2007. IPCC 4 at the RGS. Presentation at "Climate Change: Impacts, Adaptation, and Vulnerability." Royal Geographical Society, Zurich, 18 September 2007.

Dallmeier, F. 1992. "Long-Term Monitoring of Biological Diversity in Tropical Forest Areas. Methods for Establishment and Inventory of Permanent Plots." MAB Digest Series, 11, UNESCO, Paris.

Dallmeier, F., and J. A. Comiskey (eds.). 1998. Forest Biodiversity in North, Central and South America and the Caribbean: Research and Monitoring. Man and the Biosphere Series, Vol. 21. Carnforth, Lancashire, UK: UNESCO and Parthenon Publishing.

Environment Canada. 2003. EMAN: Monitoring Biodiversity in Canadian Forests. Report prepared by EMAN Coordinating Office, Burlington, Ontario, Canada.

Ervin, J., and J. Parrish. 2006. Toward a Framework for Conducting Ecoregional Threat Assessments. USDA Forest Service Proceedings, PMRS-P-42CD, pp. 105–112.

Eversham, E. M., and N. V. L. Brokaw. 1996. Forest Damage and Recovery from Catastrophic Wind. *Botanical Review*, 62(2):113–185.

Fenech, A., M. Murphy, D. MacIver, H. Auld, and R. Bing Rong. 2005. The Americas: Building the Adaptive Capacity to Global Environmental Change. Occasional Paper 5, Adaptation and Impacts Research (AIRD), Environment Canada.

Fischlin, A. 2007. Impacts on the Global Resource Base: Ecosystems. Presentation at "Climate Change: Impacts, Adaptation, and Vulnerability," IPCC, Assessment Report Four, Working Group II. Royal Geographical Society, Zurich, 18 September 2007.

Folke, C., S. Carpenter, B. Walker, M. Scheffer, T. Elmqvist, L. Gunderson, and C. S. Holling. 2004. Regime Shifts, Resilience, and Biodiversity in Ecosystem Management. *Annual Review of Ecology, Evolution, and Systematics*, 35:557–581.

Githaiga-Mwicigi, J. M. W., D. H. K. Fairbanks, and G. Midgley. 2002. Hierarchical Process Define Spatial Pattern of Avian Assemblages Restricted and Endemic to the Arid Karoo. *South African Journal of Biogeography*, 29(8):1067–1087.

Gough, W. A., and A. Leung. 2002. Nature and Fate of Hudson Bay Permafrost. *Regional Environmental Change*, 2:177–184.

Groombridge, B., ed. 1992. *Global Biodiversity: Status of the Earth's Living Resources*. London: Chapman and Hall.

Harrison, R. D. 2000. Repercussions of El Nino: Drought Causes Extinctions and the Breakdown of Mutualism in Borneo. *Proceedings of the Royal Society of London, Series B: Biological Sciences*, 267(1446):911–915.

IMAGE 1. Spatio-temporal distribution of bird species richness within the Argentine pampas agroecosystems from 2002 to 2007.

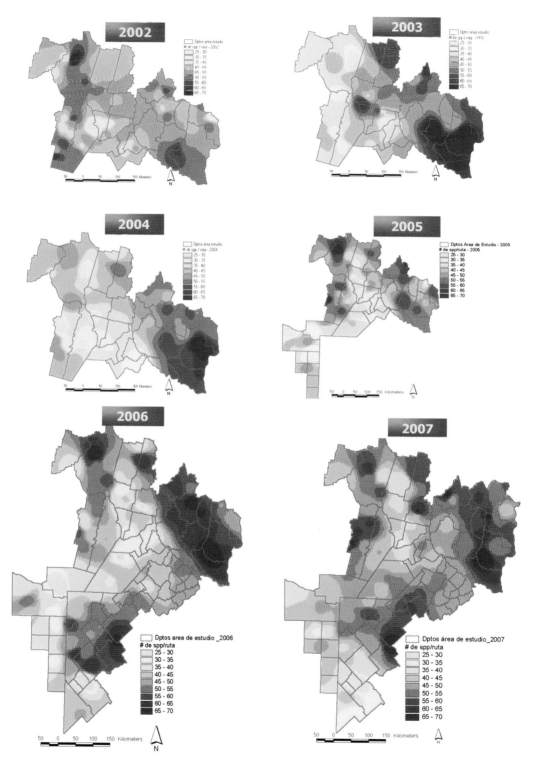

IMAGE 2. Spatio-temporal distribution of relative abundance of Swainson's Hawk within the Argentine pampas agroecosystem monitored from 2002 to 2007.

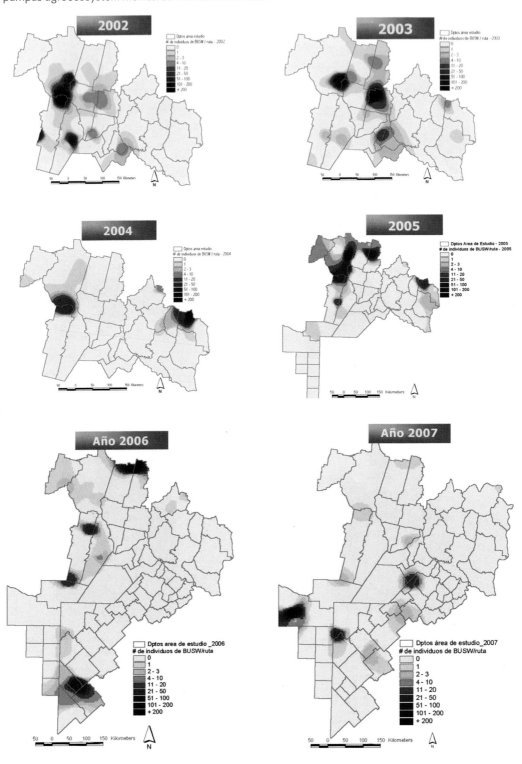

IMAGE 3. Spatio-temporal distribution of relative abundance of Burrowing Owl within the Argentine pampas agroecosystem from 2002 to 2007.

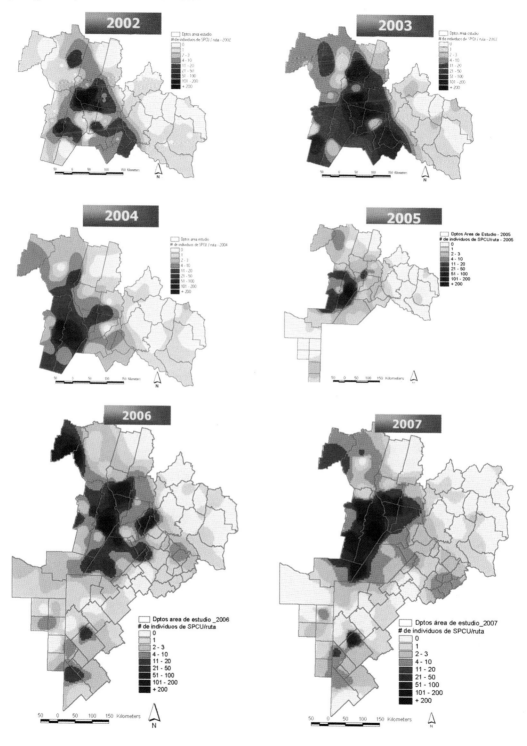

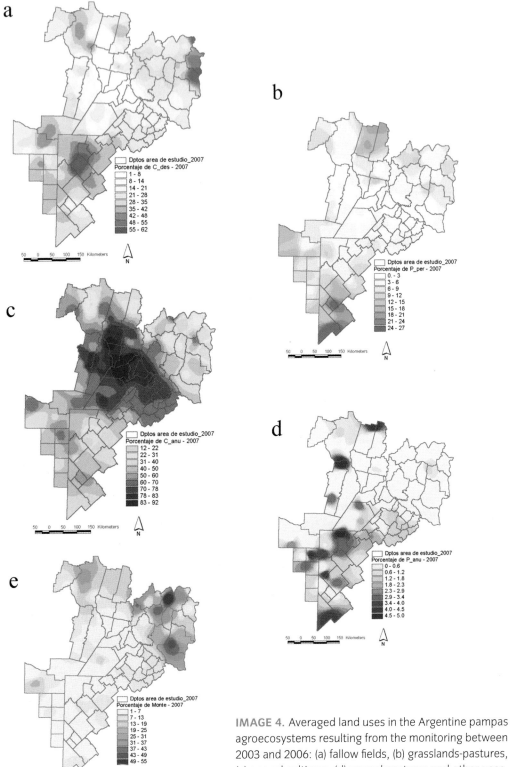

IMAGE 4. Averaged land uses in the Argentine pampas agroecosystems resulting from the monitoring between 2003 and 2006: (a) fallow fields, (b) grasslands-pastures, (c) annual cultivars, (d) annual pastures and other uses, (e) woodlands.

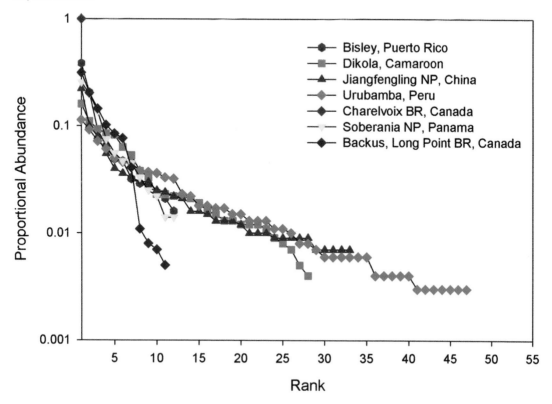

IMAGE 5. Diversity in Smithsonian International Biodiversity Monitoring Sites for Asia, Africa, South America, and Canada.

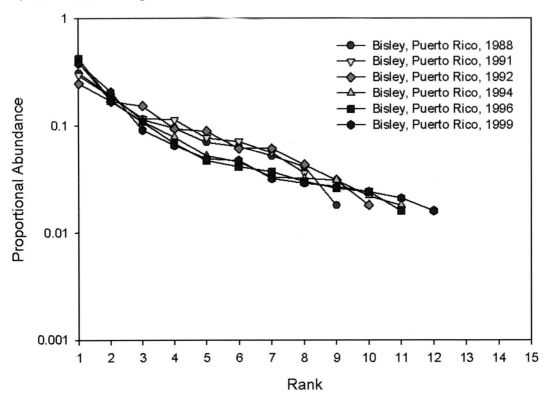

IMAGE 6. Diversity of tree species in Bisley, Puerto Rico, one year prior to Hurricane Hugo damage in 1988 and at subsequent measurement periods from 1991 to 1999. Data currently not available for more than ten top families ranked from highest to least abundance.

IMAGE 7. (a) Diversity of sugar maple site at Gananoque, Ontario, Canada, impacted by ice-storm damage in comparison to Carolinian sites at Long Point, Ontario, Canada and (b) Diversity of sugar maple at Mont. St. Hilaire, Quebec, Canada, ice storm impacted site in comparison to maple mixed wood site in Quebec, Canada.

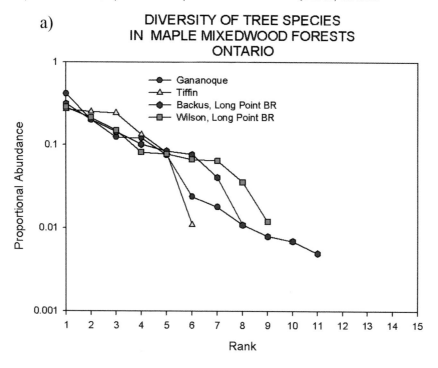

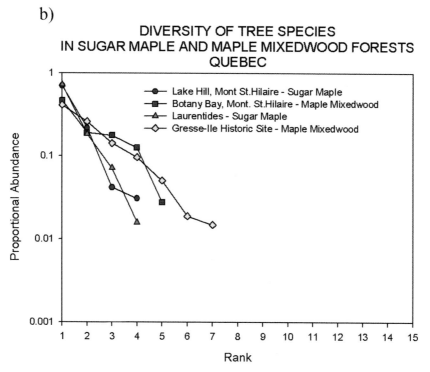

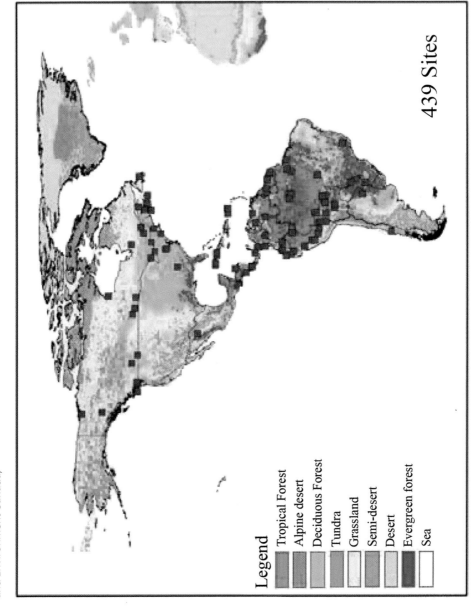

IMAGE 8. Smithsonian Institution Forest Biodiversity Observing Sites in the Americas. (Source: Smithsonian Institution and Environment Canada)

Legend

Tropical Forest
Alpine desert
Deciduous Forest
Tundra
Grassland
Semi-desert
Desert
Evergreen forest
Sea

439 Sites

IMAGE 9. Climate and forest biodiversity change: number of (a) species and (b) families as related to heat units in Canada compared to reference Smithsonian global biodiversity observation sites in the USA, Caribbean and South America. (Source: Karsh and MacIver, Environment Canada)

a)

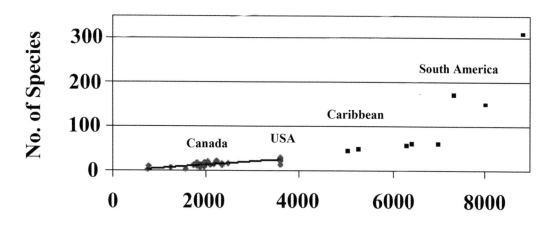

b)

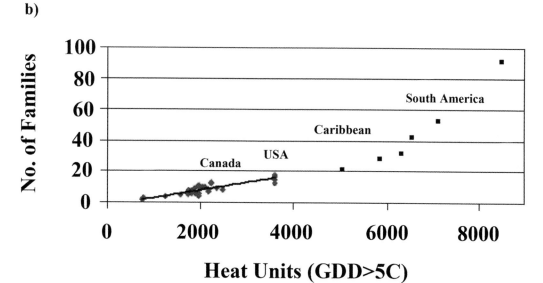

IMAGE 10. (a) Heat by family biodiversity for mixed wood forest in southern Ontario, based on Smithsonian Institution forest biodiversity observing sites, and (b) climate- based biodiversity mapping and global (SI/MAB) sites in southern Ontario. Cumulative impact of the chemical atmosphere; climate based biodiversity and Smithsonian Institution Forest Biodiversity Observing Sites. (Sources: MacIver, 1997; 1997 Canadian Acid Precipitation Assessment and the 1980–1993 Ground-Level Ozone and Its Precursors (Data) Assessment (1997), Environment Canada and Science Program)

a)

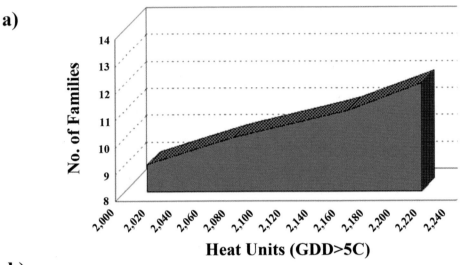

b)

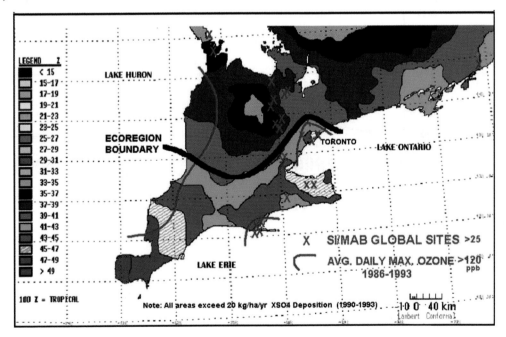

IMAGE 11. Annual record of temperature in Querétaro City for 2003, 2005 and 2006. (Source: CONAGUA, Querétaro meteorological station)

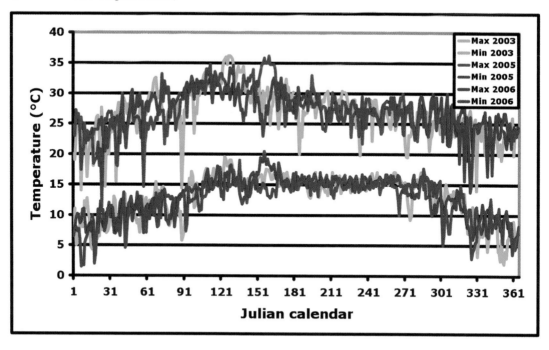

IMAGE 12. March–October precipitation in Querétaro City for 2003, 2005 and 2006. (Source: CONAGUA, Querétaro meteorological station)

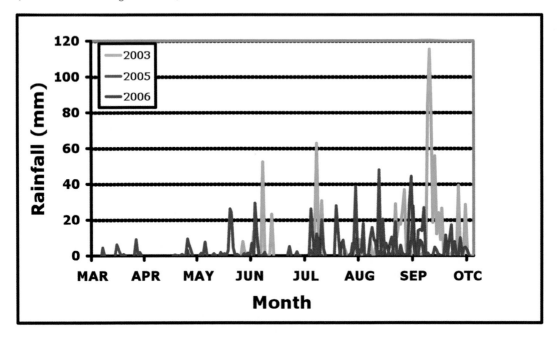

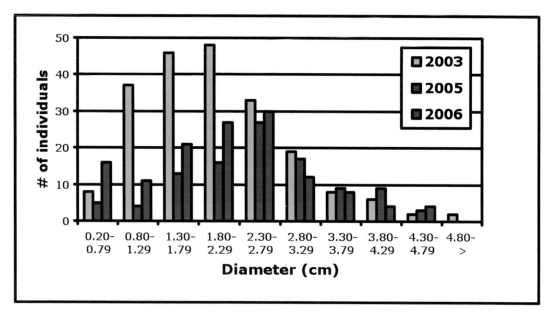

IMAGE 13. Histogram showing size–class distribution of M. mathildae individuals at Cañada Juriquilla plot. Data for 2003 were reported by Hernández-Oria et al. (2003). Total number of individuals: 2003 = 209; 2005 = 111; 2006 = 133.

IMAGE 14. Percentage of each insect order for the six focal tree species. Zero percentage indicates that the order was present for that tree in less than 0.5%. When orders are not listed, it was absent from all the samples collected for that specific tree.

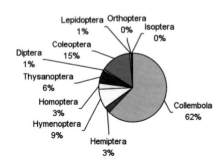

Cordia alliodora

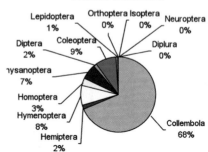

Luheia seemanii

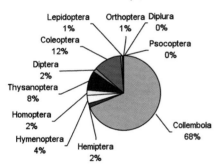

Hura crepitans

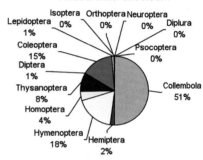

Anacardium excelsum

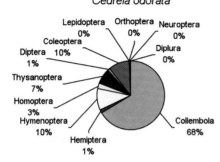

Cedrela odorata

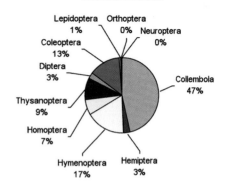

Tabebuia rosea

IMAGE 15. Percentages of litter arthropods groups (a) in monocultures and (b) in mixed species plots (3- and 6-species plots).

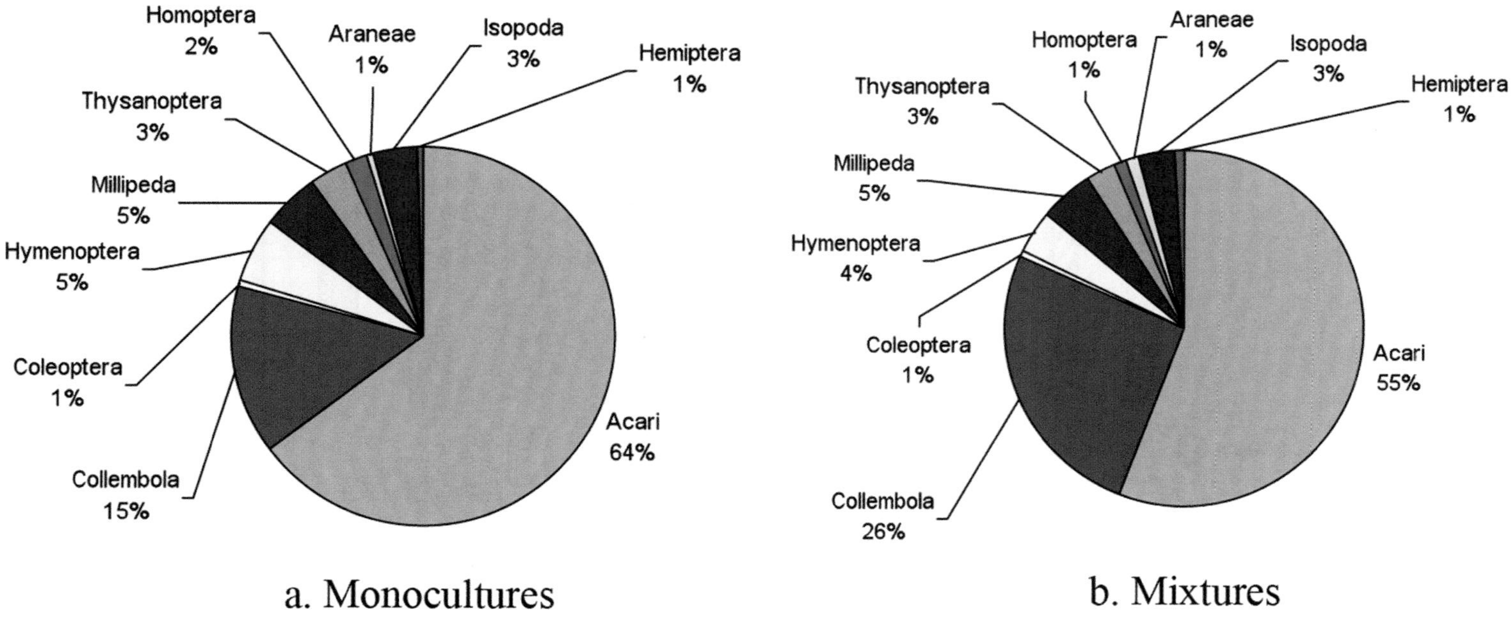

a. Monocultures

b. Mixtures

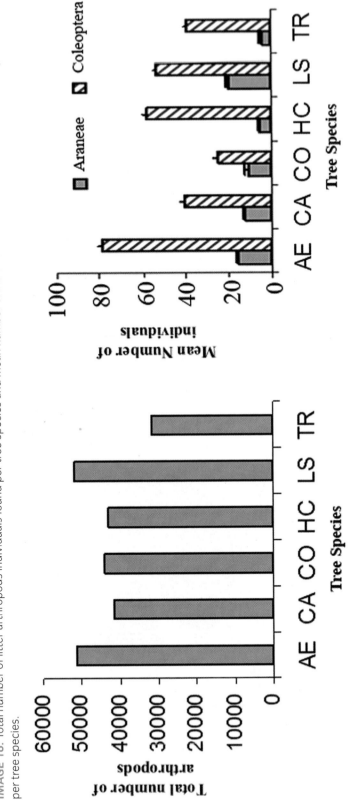

IMAGE 16. Total number of litter arthropods individuals found per tree species and mean number of individuals with SE in the *Araneae* and *Coleoptera* orders per tree species.

[IPCC] Intergovernmental Panel on Climate Change. 2001. *Climate Change 2001: Impacts, Adaptation, and Vulnerability.* Contribution of Working Group II to the Third Assessment Report of the IPCC. Cambridge, UK: Cambridge University Press.

IPCC. 2007. *The Physical Science Basis: Summary for Policymakers.* Contribution of Working Group I to the Fourth Assessment Report of the IPCC. Cambridge, UK: Cambridge University Press.

Jentsch, A., J. Kreyling, and C. Beierkuhnlein. 2007. A New Generation of Climate-Change Experiments: Events, Not Trends. *Frontiers of Ecology and the Environment*, 5(7):365–374.

Karsh, M. B., D. C. MacIver, A. Fenech, and H. Auld. 2007. Climate-Based Predictions of Forest Biodiversity Using Smithsonian's Global Earth Observing Network. Occasional Paper 8, Adaptation and Impacts Research (AIRD), Environment Canada.

Kharin, V. V., F. W. Zwiers, X. Zhang, and G. C. Hegerl. 2007. Changes in Temperature and Precipitation Extremes in IPCC Ensemble of Global Coupled Model Simulations. *Journal of Climate*, 20(8):1419–1444.

Killeen, T. J. 2007. A Perfect Storm in the Amazon Wilderness: Development and Conservation in the Context of the Initiative for the Integration of the Regional Infrastructure of South America (IIRSA). Advances in Applied Biodiversity Science, No. 7.

Krockenberger, A. K., R. L. Kitching, and S. M. Turton (eds.). 2003. Environmental Crisis: Climate Change and Terrestrial Biodiversity in Queensland. Cooperative Research Centre for Tropical Rainforest Ecology and Management. Rainforest CRC, Cairns.

Laurenroth, W. K., H. E. Epstein, J. M. Paruelo, I. C. Burke, M. R. Aguiar, and O. E. Sala. 2004. Potential Effects of Climate Change on the Temperate Zones of North and South America. *Revista Chilena de Historia Natural*, 77:439–453.

Lavorel, S., M. D. Flannigan, E. F. Lambin, and M. C. Scholes. 2007. Vulnerability of Land Systems to Fire: Interactions among Humans, Climate, the Atmosphere, and Ecosystems. *Mitigation and Adaptation Strategies for Global Change*, 12(1):33–53.

Leemans, R., and A. van Vliet. 2005. Responses of Species to Changes in Climate Determine Climate Protection Targets. Environmental Systems Analysis Group, Wageningen, The Netherlands.

MacIver, D. 1998. Atmospheric Change and Biodiversity. *Environmental Monitoring and Assessment*, 49:177–189.

Malcolm, J. R., C. Liu, R. P. Neilson, L. Hansen, and L. Hannah. 2006. Global Warming and Extinctions of Endemic Species from Biodiversity Hotspots. *Conservation Biology*, 20(2):538–548.

Markham, A. 1996. Potential Impacts of Climate Change on Ecosystems: A Review of Implications for Policymakers and Conservation Biologists. *Climate Research*, 6:179–191.

McCready, J. 2004. Ice Storm 1998: Lessons Learned. 6th Canadian Urban Forest Conference, Kelowna, British Columbia.

[MEA] Millennium Ecosystem Assessment. 2005. *Ecosystems and Human Well-Being: Biodiversity Synthesis.* Washington, DC: World Resources Institute.

Miles, L., A. Grainger, and O. Phillips. 2004. The Impact of Global Climate Change on Tropical Forest Biodiversity in Amazonia. *Global Ecology and Biogeography*, 13:553–565.

Miller, M. 2006. The San Diego Declaration on Climate Change and Fire Management: The Association of Fire Ecology. Third International Fire Ecology and Management Congress, San Diego, California.

Nielson, C., O. Van Dyke, and J. Pedlar. 2003. Effects of Past Management on Ice Storm Damage in Hardwood Stands in Eastern Ontario. *Forestry Chronicle*, 79(1):70–74.

Nobre, C. A. 2007. Climate Change Scenarios and Impacts on the Biomes of South America. Presentation at Centro de Previsão de Tempo e Estudos Climáticos (CPTEC), Instituto Nacional de Pesquisas Espaciais (INPE), Cachoeira Paulista, SP-Brazil.

Odion, D. C., E. J. Frost, J. R. Stritthold, H. Jiang, D. A. Dellasala, and M. A. Moritz. 2004. Patterns of Fire Severity and Forest Conditions in the Western Klamath Mountains, California. *Conservation Biology*, 18(4):927–936.

Olson, D. M., E. Dinerstein, E. D. Wikramanayake, N. D. Burgess, G. V. N. Powell, E. C. Underwood, J. A. D'Amico, I. Itoua, H. E. Strand, J. C. Morrison, J. Loucks, T. F. Allnutt, T. H. Ricketts, Y. Kura, J. F. Lamoreux, W. W. Wettengel, P. Hedao, and K. R. Kassem. 2001. Terrestrial Ecoregions of the World: A New Map of Life on Earth. *Bioscience*, 51(11):933–938.

Peters, D. P. C., R. A. Pielke, B. T. Bestelmeyer, C. D. Allen, S. Munson-McGee, and K. M. Havstad. 2004. Cross-scale Interactions, Nonlinearities, and Forecasting Catastrophic Events. *Proceedings of the National Academy of Sciences*, 101(42):15130–15135.

Reusch, T. B. H., A. Ehlers, A. Hammerli, and B. Worm. 2005. Ecosystem Recovery after Climatic Extremes Enhanced by Genotypic Diversity. *Proceedings of the National Academy of Sciences of the United States of America*, 102(8):2826–2831.

Rochefort, L., and F. I. Woodward. 1992. Effects of Climate Change and a Doubling of CO_2 on Vegetation Diversity. *Journal of Experimental Botany*, 43:1169–1180.

Stocks, B. J., M. A. Fosberg, T. J. Lynham, L. Mearns, B. M. Wotton, Q. Yang, J. Z. Jin, K. Lawrence, G. R. Hartley, J. A. Mason, and D. W. McKenney. 1998. Climate Change and Forest Fire Potential in Russian and Canadian Boreal Forests. *Climate Change*, 38:1–13.

Stott, A. 2007. Biodiversity-Climate Interactions: Adaptation, Mitigation and Human Livelihoods. A Policy Perspective. Presentation at "Biodiversity-Climate Interactions." Royal Society, London, 12 June 2007.

Szaro, R. C., and B. K. Williams. 2008. "Climate Change: Environmental Effects and Management Adaptations." In *Climate Change and Biodiversity in the Americas*, ed. A. D. Fenech, D. MacIver, and F. Dallmeier, pp. 277–294. Toronto, ON: Environment Canada.

Thomas, C. D., A. Cameron, R. E. Green, M. Bakkenes, L. J. Beaumont, Y. C. Collingham, B. F. N. Erasmus, M. F. de Siqueira, A. Grainger, L. Hannah, L. Hughes, B. Huntley, A. S. van Jaarsveld, G. F. Midgely, L. Miles, M. A. Ortega-Huerta, A. Townsend Peterson, O. L. Phillips, and S. E. Williams. 2004. Extinction Risk from Climate Change. *Nature*, 427:145–148.

van Bloem, S. J., A. E. Lugo, and P. G. Murphy. 2006. Structural Response of Caribbean Dry Forests to Hurricane Winds: A Case Study from Guanica Forest, Puerto Rico. *Journal of Biogeography*, 33(3):517–523.

Varrin, R., J. Bowman, and P. A. Gray. 2007. The Known and Potential Impacts of Climate Change on Biodiversity in Ontario's Terrestrial Ecosystems: Case Studies and Recommendations for Adaptation. Climate Change Research Report CCRR-09, Applied Research and Development Branch, Ontario Ministry of Natural Resources, Sault Ste. Marie, Ontario.

Wilson, J. B., and A. D. Agnew. 1992. Positive Feedback Switches in Plant Communities. *Advances in Ecological Research*, 23:263–336.

WWF. 2007. Ecoregions: Terrestrial Ecoregions. Available at http://www.worldwildlife.org/science/ecoregions/item1267.html (accessed 26 April 2009).

Phenological Changes of *Mammillaria mathildae* Associated to Climatic Change in a Deciduous Tropical Forest

Oscar García-Rubio[1] and Guadalupe Malda-Barrera[1]

ABSTRACT: Phenological phases of the endemic cactus *Mammillaria matlidae* were recorded as part of the long-term monitoring program for natural protected areas in a deciduous tropical forest near Querétaro City. Commonly this cactus blooms and fruits from late May to August. However the blooming period in 2005 and 2006 occurred one and a half months earlier. Since phenophases depend on seasonal meteorological trends, their association with a temperature increase was explored. Nonsignificant differences were found between 2003, 2005 and 2006. In contrast, rainfall patterns were different and the rainy season was atypical in 2003: it began later, was shorter and extreme. In just the first nine days of October precipitation registered 432 mm, accumulating more water than 2005 total precipitation (405 mm). Therefore it was the highest precipitation in the last 70 years. This negatively impacted both the number of *Mammillaria matlidae* individuals (decreasing from 209 to 111) and the cactus's annual recruitment. Meanwhile, during 2005 and 2006 rainfall patterns were homogeneously distributed throughout the year. In both years *M. mathildae* fruit yield increased and plant population increased in 2006, with 31 seedlings registered. It is possible that one of the key factors for the plant population recovery was the homogeneous distribution of rainfall. These observations contribute to a better understanding of climate change

[1] Universidad Autónoma de Querétaro, Laboratory of Plant Physiology, Av. de las Ciencias S/N, Juriquilla Querétaro, Santiago de Querétaro, Mexico C.P. 76230.

Corresponding author: O. García-Rubio (osrigaru@gmail.com)

repercussions on individual species, and further, they could facilitate improvement of the construction of prediction models to estimate potential distribution of species after climate change events.

Keywords: *climate change, recruitment, Mammillaria mathildae, phenology*

INTRODUCTION

For the past decades the evidence of global climate change as a consequence of the increased emission of greenhouse effect gases has been overwhelming (Joyce et al., 2006). One of the Intergovernmental Panel on Climate Change's (IPCC) conclusions states that "it is at least 90 percent certain that temperatures will continue to rise, as average global surface temperature is projected to increase by between 1.4 and 5.8°C above 1990 levels by 2100" (Gucinski et al., 2004). It is evident that anthropogenic greenhouse gas emissions will continue to affect us for a long time. Even with an ideal scenario where CO_2 emissions stopped today, the reestablishment of atmospheric CO_2 concentrations to preindustrial times would take many centuries!

The consequences of climate change (for example, increases in extreme weather events, displacement of high-latitude species by low-latitude ones, or impacts observed on climate-sensitive resources for earlier phenology events) will affect global ecosystems. Therefore, it is necessary to predict the possible changes and develop contingency plans for alternative scenarios. Some of the crucial challenges could be: modeling how ecosystems will respond to climate change, how would climate change interact with other stressors (for example, invasive species, urban expansion, agriculture and livestock), or how to maintain the sustainability of natural areas.

However regional and local scales consequences have been difficult to predict due to a synergistic effect of the stressors. Developing strategies to deal with such vulnerabilities requires the integration of information at multiple scales. In this context, monitoring sensitive populations could be configured to identify alternative conditions as they may occur in order to contribute to the development of further accurate prediction models.

Mexico emits per capita 0.96 tons of carbon (C) per year to the atmosphere (Martínez and Fernández, 2004), and 30.5% of these emissions are directly related to land-use changes (urbanization, agriculture and livestock). It is estimated that temperate forests, tropical forests and high mountain mesophyllous forests will be the most threatened ecosystems in Mexico. In contrast, perennial tropical forests, deciduous and semideciduous tropical forests will increase their distribution range (Arriaga and Gómez, 2004).

Some relics of deciduous tropical forest are distributed around Querétaro City in central Mexico, which Arriaga and Gómez (2004) predict are very susceptible to radiate in a wider range as a consequence of climate change. For this reason, these particular ecosystems are good models for monitoring climate changes, as well as to determine how the vegetation will modify its abiotic and biotic interactions as a means of predicting how they will fit in the new environments that may result.

In recent years, Querétaro City has experienced its most extreme rainfall in 70 years (e.g., one day the rainfall measured 115.2 mm). In 2003 the annual precipitation reached 1,018 mm whereas in 2005 and 2006 it was 404 mm and 645 mm, respectively. These extreme events changed the vision of climate change, and some researchers began to associate the unusual vegetation phenology with the rare climate events. Since phenological rhythms of plant species depend on seasonal meteorological trends, these phenophases can be used to monitor climate change. Therefore, in this research, the immediate goal was to assess if early flowering events observed for *Mammillaria mathildae*, an endemic cactus associated to deciduous tropical forest in central Mexico, are related to temperature increases and/or to changes in precipitation patterns. Also how these climate events impact on the cactus's annual recruitment was examined. In this way, phytophenology could provide some indirect information about how the environmental changes impact on individual species.

METHODS

Study Area

"Cañada Juriquilla" is located north of Querétaro City in the province of Juriquilla, at an altitude of 1,890 m, between 20°41′35.2″ N and 100°27′16.4″ W. It includes the largest population of *M. mathildae* (133 individuals in 2006). "La Cañada" is located southwest of Querétaro City in the province of Corregidora, at an altitude of 1,870 m, between 20°37′31.8″ N and 100°18′52.0″ W (Cabrera and Gómez, 2003).

Climate Data

Daily temperature and precipitation data for the study area were obtained from CONAGUA's National Ministry of Water in Querétaro meteorological station (20°35′ N: 100°24′ W).

Flowering and Fruit Set Evaluation

Periodic surveys were performed in order to determine the beginning and end of blooming and fruiting events. *M. mathildae* plant size was positively correlated with flowering ratio; therefore only 1.8–5 cm plants were considered. Phenology was determined in situ by direct observation from early February to late October. The number of flowers per individual was counted, following fruit yield determination. Reproductive effort (RE) was calculated for each year as a flowers/fruits ratio.

Seed Collection

Fruits from 20 different donor plants were collected from Cañada Juriquilla's population during August 2005 and 2006, following recommendations described by Ross (2004). Once in the laboratory, seeds were manually extracted and counted.

Population Census

The census of the Cañada Juriquilla's cactus population was performed in situ by direct observation in late December 2005 and early January in both 2005 and 2006. Each individual cactus was measured (diameter, height and number of areoles) and systematically tagged.

Statistical Analysis

A single factor ANOVA was used to test flower and fruit yield differences, and to compare temperature patterns, at probability $p < 0.05$. Statistical analyses were performed using JMP 6.0 statistical software (SAS Institute, Inc., 2005).

RESULTS AND DISCUSSION

Phenological Gap

It is important to note that different *M. mathildae* populations seem to be moderately well synchronized in their phenology. For the five known described populations (García and Malda-Barrera, 2007), blooming periods coincided in all cases. According to Cabrera and Gómez (2003), La Cañada's population bloomed from late May to late June and bore fruit from June to August in 2003; whereas Hernández-Oria et al. (2003) registered a similar response for Cañada Juriquilla's population in 2001 (flowering recorded in May).

However, in both 2005 and 2006, the first flowering events appeared earlier, during the first days of April, continuing until late June, and the maximum blooming rate was reached from early to middle May. Meanwhile, the fruit season began in early May and extended until late July.

This phenological change (one and a half month earlier) can be attributed to many reasons since many biochemical and physiological processes are highly correlated with temperature (Taiz and Zeiger, 2002). Early flowering has been reported around the world on account of climate change. For instance, in Europe, the initiation of phenological phases (leaf formation and flowering) has occurred earlier in many species as a consequence of the rise in Spring temperatures (Schwartz, 1999). Parmesan and Yohe (2003) indicated that phenological changes in African flora are a response to global warming. Weiss and Overpeck (2005) presented data showing widespread warming trends in winter and spring, decreased frequency of freezing temperatures, lengthening of the freeze-free season, and increased minimum winter temperatures. All of these factors affected phenological events over the Sonoran Desert vegetation.

In this study, the annual temperature patterns observed for the study area during 2003, 2005 and 2006 were not significantly different: $F = 0.1903$ Prob. $> F < 0.9024$ (Image 11, see photospread). Some climate variables other than temperature could be related to the observed changes in blooming periods. Vegetation phenology in arid and semiarid ecosystems and seasonally dry tropical climates (as in deciduous tropical forests) is controlled primarily by water availability (Justiniano and Fredericksen, 2000). As global climate change can modify rainfall seasonality, extreme weather events like floods or droughts, as well as longer dry or rainy seasons, can occur (Fauchereau et al., 2003).

Comparing registered precipitation in 2003, 2005 and 2006, the first important 2003 precipitation occurred the last week of May. In contrast, 2005 and 2006 rainfall events were recorded starting in March (Image 12, see photospread). These early rainfalls could be associated with the early blooming and fruit seasons observed for *M. mathildae*. Thus, water availability allowed cacti to initiate their annual growth period, and developed flowers earlier compared to 2003, when the blooming started in late May.

Mammillaria mathildae Recruitment

Precipitation occurred more uniformly in 2005 and 2006 than in 2003 when rain was less frequent but unusually intense (Image 12). In order to distinguish some possible effects of such extreme climate events on *M. mathildae* populations, a fruit yield comparison for different years (Table 1) was conducted. In 2003, through the late rainfall season, fruit production was very poor in contrast to 2005 and 2006. Furthermore, the population of *M. mathildae* decreased in number of individuals; declining from 209 to 111 plants (and only five seedlings were registered) in 2005 (Image 13, see photospread).

A possible explanation for this population loss could be that excessive amounts of water negatively affected young individuals. Image 13 shows that small plants reduced their number, while larger plants persisted in the field. In addition, the heavy 2003 precipitation affected the 2005 recruitment season; probably due to excess of running water that could have damaged or removed plantlets and seeds from the soil seed bank (*M. mathildae* occurs in pronounced slopes), therefore limiting future recruitment periods. Some other *M. mathildae* germination experiments in which seeds were germinated after emulating a running water event (15 minutes) showed a seed viability loss from 70% (control) to 32% in the treated seeds (García and Malda-Barrera, 2006). Also in situ recruitment for the 2003 period was very low (Image 13). There is no evidence of actively growing populations since medium-sized cactus were the more abundant (Hernández-Oria et al., 2003).

In contrast, the homogeneous rain distribution in 2005 and 2006 (Image 12) resulted in a population increase (Image 13, see photospread); 133 individuals with 31 seedlings recorded in 2006. In 2005, March precipitation stimulated the early floral production (14.5 ± 4.34 flowers). In 2006, the late March rainfall triggered flower emergency, which reached 16.0 ± 5.22 flowers, similar to the 2005 period (F = 2.1992 Prob. > F < 0.1463). Nevertheless, fruit development performed differently, because in contrast to 2005, the 2006 constant precipitation beginning in late April promoted a yield of 9.0 ± 4.58 fruits,

TABLE 1. Quantitative phenology of *M. mathildae* population.

Year	No. Flowers (mean ± SE[a])	No. Fruits (mean ± SE[a])	Reproductive Effort
2003[b]	31	2	0.15
2005[c]	14.5 ± 4.34	6.5 ± 2.87	0.47 ± 0.13
2006[c]	16.0 ± 5.22	9.0 ± 4.58	0.59 ± 0.14

[a] SE = Standard Error.
[b] Data from Hernández-Oria et al. (2003).
[c] 2005 and 2006 data are the mean ± SE for 20 individuals.

which was significantly different from 6.5 ± 2.87 yield registered in 2005 (F = 7.4602 Prob. > F < 0.0095). It is probable that the absence of rain for five weeks in 2005 had a negative impact on fruit production; compromising the population RE (see Table 1) which was also significantly different (F = 7.7186 Prob. > F < 0.0084). Recent precipitation (2007) presented a similar rain pattern to 2006 (data not shown). This fact combined with a higher number of fruits produced, could result in a seedling number increase compared to 2006 recruitment period.

CONCLUSION

Evidently, there are long-term climate phenomena in central Mexico that require explanation and understanding. A major future challenge is to better understand which systems or species are most or least susceptible to climate change scenarios. There are different approaches to develop prediction models. For example, some researchers like Trenberth et al. (2003) emphasized that major focus must be placed on precipitation amount, rather than other precipitation parameters that also may change (frequency, timing, intensity, statistical distributions, extremes, types of events, etc.). However, this study illustrates that annual rain distribution influences seedling recruitment affecting population dynamics of *M. mathildae*. De Steven and Wright (2002) found that recruitment in tropical trees was synchronized with seed production, and thus also appeared to be partly influenced by El Niño climate events. As a consequence, seedling recruitment varied temporally and spatially. Since different species have distinct phenophases, it is necessary to implement long-term studies for other species as well, particularly "nurse trees," which are important to understand the vegetation dynamics in deciduous tropical forests.

Such observations contribute to a better understanding of these relations and therefore facilitate the improvement for the construction of prediction models to estimate species potential distribution after any climate event. Obviously many other topics need to be reviewed, like precipitation abundance that could be related to the increase in vegetal biomass, which in the dry season will lead to higher fire frequencies. Fires discourage deciduous tropical forest recovery solidifying the competitive advantage of invasive species such as rose Natal grass (*Melinis repens*) (Martínez and Fernández, 2004).

ACKNOWLEDGMENTS

The authors are indebted to Ms. Teresa Rubio for her technical assistance in flower and fruit counting. We thank FIQMA for providing access to Cañada Juriquilla Natural Protected Area. We thank Idea Wild for equipment donation for this project. This work was partially funded by CONACYT (Project CONACYT-CONAFOR-2004-C01-71).

REFERENCES

Arriaga, L., and L. Gómez. 2004. "Posibles efectos del cambio climático en algunos componentes de la biodiversidad de México." In *Cambio Climático: una visión desde México*, ed. J. Martínez and B. Fernández, pp. 255–265. México, D.F.: Instituto Nacional de Ecología (INE-Semarnat).

Cabrera, L. J. A., and S. M. Gómez. 2003. Algunos aspectos de la biología floral de *Mammillaria mathildae*, especie microendémica del estado de Querétaro y en peligro de extinción. Memorias del Quinto Verano de la Ciencia de la Región Centro y Segundo Verano de la Ciencia de la UAQ, Querétaro, Mex. UAQ 50.

De Steven, D., and J. Wright. 2002. Consequences of Variable Reproduction for Seedling Recruitment in Three Neotropical Tree Species. *Ecology,* 83:2315–2327.

Fauchereau, N., M. Trzaska, and Y. Richard. 2003. Rainfall Variability and Changes in Southern Africa during the 20th Century in the Global Warming Context. *Natural Hazards,* 29:139–154.

García-Rubio, O., and G. Malda-Barrera. 2006. Factores que afectan la germinación de *Mammillaria mathildae* (Krähenbühl et Krainz, 1973). IX Congreso Latinoamericano de Botánica, Santo Domingo, República Dominicana, 18–25 June 2006.

———. 2007. A Rapid Methodology to Select Urban Conservation Areas. A Case of Study: *Mammillaria mathildae* Conservation. Society for Conservation Biology 21st Annual Meeting, Port Elizabeth, South Africa, 1–5 July 2007.

Gucinski, H., R. P. Neilson, and S. McNulty. 2004. "Implications of Global Climate Change for Southern Forests: Can We Separate Fact from Fiction?" In *Southern Forest Science: Past, Present and Future,* ed. M. Rauscher and K. Johnson, pp. 365–371. Gen. Tech. Rep. GTR-SRS-75. Asheville, NC: U.S. Department of Agriculture, Forest Service, Southern Research Station.

Hernández-Oria, J. G., R. Chávez, M. Galindo, G. Hernández, M. Lagunas, S. Martínez, R. Mendoza, A. Sánchez, and M. Sánchez. 2003. Evaluación de Aspectos Ecológicos de una Nueva Población de *Mammillaria mathildae* Kraehenbuehl & Krainz en Querétaro. *Cactáceas y Suculentas Mexicanas,* 47:4–10.

Joyce, L., R. Haynes, R. White, and R. J. Barbour (tech. coords.). 2006. *Bringing Climate Change into Natural Resource Management: Proceedings.* Gen. Tech. Rep. PNW-GTR-706. Portland, OR: U.S. Department of Agriculture, Forest Service, Pacific Northwest Research Station.

Justiniano, M. J., and T. S. Fredericksen. 2000. Phenology of Tree Species in Bolivian Dry Season. *Biotropica,* 32:276–281.

Krähenbühl, F., and H. Krainz. 1973. *Mammillaria mathildae. Kakteen Sukkulenten,* 24:265.

Martínez, J., and B. Fernández. 2004. *Cambio Climático: una visión desde México.* México, D.F.: Instituto Nacional de Ecología (INE-Semarnat).

Parmesan, C., and G. Yohe. 2003. A Globally Coherent Fingerprint of Climate Change Impacts across Natural Systems. *Nature,* 421:37–42.

Ross, C., 2004. Native Seed Collection and Use in Arid Land Reclamation: A Low-Tech Approach. *Environmental Monitoring and Assessment,* 99:1–3.

SAS Institute. 2005. JMP 6.0 statistical software. Cary, NC.

Schwartz, M. D. 1999. Advancing to Full Bloom: Planning Phenological Research for the 21st Century. *International Journal of Biometeorology,* 42:113–118.

Taiz, L., and I. Zeiger. 2002. *Plant Physiology.* Sunderland, MA: Sinauer Associates.

Trenberth, K. E., A. Dai, R. M. Rasmussen, and D. B. Parsons. 2003. The Changing Character of Precipitation. *Bulletin of the American Meteorological Society,* 84:1205–1217.

Weiss, J. L., and J. T. Overpeck, 2005. Is the Sonoran Desert Losing Its Cool? *Global Change Biology,* 11:2065–2077.

Tropical Tree Plantations with Native Species

Linking Carbon Storage with Concerns for Biodiversity

Malena Sarlo,[1,2] Chrystal Healy[1,3] and Catherine Potvin[1,4]

ABSTRACT: Increased concern for climate change coupled with a willingness of some developed countries to adopt mitigation strategies has raised the interest for tropical tree plantations as a possible option. Yet the expansion of monocultures raises significant concern for biodiversity. Here, the extent to which tree diversity affects plot productivity, carbon storage and, in turn, earthworm numbers and soil and litter arthropod diversity is examined. The study was carried out in a tropical tree plantation located in Sardinilla, Panama, where six native trees have been planted in different diversity treatments to test for the effects of biodiversity on ecosystem functioning. ANOVAR unveiled that the tree biomass of the three-species plots was significantly higher than that of the other diversity levels and this effect increased through time. Espave (*Anacardium excelsum*) had the smallest average biomass while Cedro amargo (*Cedrela odorata*) was the most productive species in the three- and six-species mixtures. The additive partitioning method used to compare the basal area of mixtures with monocultures demonstrated a significant positive effect of

[1]*Department of Biology, McGill University, 1205 Docteur Penfield Avenue, Room W6/8, Montréal H3A 1B1, Québec, Canada.*

[2]*The Nature Conservancy—Panamá Program, #PTY 8149, 1601 NW 97th Ave, P.O. Box 025207, Miami, FL 33102-5207 USA.*

[3]*Quebecor World Inc., Environment, 999 de Maisonneuve Blvd., West Montréal H3A 3L4, Québec, Canada.*

[4]*Smithsonian Tropical Research Institute, Panama, Republica de Panamá.*

Corresponding author: M. Sarlo (msarlo@tnc.org).

complementarity (t = 2.64, p = 0.023) and a significant negative effect of selection (t = −2.21, p = 0.049). The highest carbon storage (5.87 tC ha⁻¹) after four years of growth corresponded to a three-species plot planted with Cedro amargo, Tronador (*Hura. crepitans*) and Ouasimo Colorado (*Luhea seemannii*), whereas, the lowest value was that of the Espave monocultures (<1 tC ha⁻¹). The effect of mixture (mono-culture vs. mixed species pairs) on the number of litter arthropods was statistically significant (Pillai Trace value = 0.130, p = 0.040), fewer arthropods were found below trees growing in monocultures than in mixed species pairs. Shannon's and Simpson's indices of diversity revealed that mixed-species pairs sustain less diverse arthropod communities than monoculture pairs. MANOVA unveiled a significant effect of tree species (Pillai Trace value = 0.647, p = 0.014) on the numbers of the litter arthropods. Two taxa, *Araneae* and *Coleoptera*, drove the significant response to tree species identity (respectively $F_{5,108} = 5.317$, p = 0.0001; $F_{5,108} = 2.324$, p = 0.048). The tree species main effect was statistically significant ($F_{5,91} = 3.28$, p = 0.009) and explained 28% of the variation in earthworm numbers. Overall, the study suggests that reforestation strategies with native tree species plantations translated into significant biodiversity benefits with correlated impacts on ecosystem carbon cycling.

Keywords: *tropical trees, biodiversity plantation, ecosystem function, arthropod diversity*

INTRODUCTION

In the wake of the Kyoto Protocol, the social and economic benefits of carbon (C) sequestration have launched tropical tree plantations to the forefront of possible mitigation strategies (Bristow et al., 2006; Cyranoski, 2007; Fonseca et al., this volume). Research has demonstrated that the efficiency by which plantations sequester C depends upon their adequate design and management and that species choice is of fundamental importance (Erskine et al., 2006; Grant, 2006; Montagnini and Porras, 1998; Redondo-Brenes and Montagnini, 2006). With the total area of plantation forests increasing rapidly (Carnus et al., 2006), it is clear that a better understanding of the long-term effects of diversity in tree plantations is needed (Vieira et al., 2004). The Sardinilla plantation in Panama, where the present study was carried out, is one of four plantations worldwide specifically designed to test the relationship between diversity and ecosystem functioning (BEF) (Scherer-Lorenzen et al., 2005).

BEF research hopes to gain insight into the consequences on ecosystems of global changes in biodiversity (Naeem and Wright, 2003; Schmid et al., 2002; Singh, 2002; Srivastava and Vellend, 2005; Tilman, 1999). While much experimental research supports the notion of a positive relationship between diversity and ecosystem function (Bauhus et al., 2000; Fridley, 2003; Lanta and Leps, 2006; Loreau and Hector, 2001; Tilman, 1999) other studies failed to uncover such a relationship (Cardinale, 2006; Wardle et al., 1996). Apparently, the inconsistencies in the literature are context dependent, resulting from the different systems and the number of trophic levels being studied, and due to discrepancies within experimental methods, including the type of traits being measured (Schlapfer and Schmid, 1999; Srivastava and Vellend, 2005).

Amongst the limitations of BEF research to date, it was noted that most experimental work has been conducted over short time periods and at very small spatial scales, measuring the same ecosystem processes, with a bias toward plants and grasslands (Carnus et al., 2006; Naeem and Wright, 2003; Schlapfer and Schmid, 1999). These experiments have altered only species richness, while keeping evenness constant and spatial and environmental heterogeneity uncontrolled (Loreau and Hector, 2001). The importance of considering the different components of diversity, such as species abundance, was illustrated by Amoroso and Turnblom (2006) who studied the differences existing between pure and mixed species plots at different densities and found that mixtures only displayed an increase in productivity at higher densities. Finally, the first generation of BEF experiments focused largely on traits at the plot level, while understanding of forested ecosystem demands incorporation of traits at the individual tree level (Potvin and Gotelli, 2008; Scherer-Lorenzen et al., 2005).

The BEF relationship is characterized by complexity such that biodiversity not only affects ecosystem processes, but is affected by them as well (Hooper et al., 2005). Here, the extent to which tree diversity affects, on the one hand, plot productivity thus C storage, and, on the other hand, the soil and litter fauna is examined. Arthropod diversity in the tropics surpasses that of any other terrestrial system on Earth (Erwin, 1982; Novotny et al., 2002; Vojtech and Basset, 2005; Wright, 2002). Large groups of litter and soil fauna, like earthworms, make up most of the animal biomass present in these environments (Fragoso and Lavelle, 1992) and affect the physical characteristics of the soil (through litter fragmentation, turn over of the soil, creation of soil aggregates, etc.) more than microorganisms such as bacteria and fungi. Ecosystem engineers like earthworms, termites and ants have proven to drastically change the soil conditions (Anderson, 1988; Anderson and Ingram, 1993; Jones et al., 1994; Lal, 1987) thus affecting plant performance. There are other complex interactions between soil dwellers and plants potentially affecting plants' performance. For instance, *Collembola*, which usually dominates soil communities, are fungivores (Badejo and Tian, 1999; Heneghan et al., 1998) and by selectively grazing on different groups of fungi they influence their biomass and activity in turn altering plant growth. The effect of plant diversity on soil biota is yet to be explored in depth (Wardle, 2002).

METHODS

Study Site and Experimental Design

The study was conducted in a tree plantation established over pasture land in 2001 in Sardinilla, central Panama (Scherer-Lorenzen et al., 2005). The soils are mostly clayish, with Aquic Tropudalfs making up the low lying areas and Typic Tropudalfs making up the upper slopes (Potvin et al., 2004). Covering a total of 9 ha, the plantation comprises six native species: Ouasimo colorado (*Luehea seemannii*-LS), Laurel (*Cordia alliodora*-CA), Espave (*Anacardium excelsium*-AE), Tronador (*Hura crepitans*-HC), Cedro amargo (*Cedrela odorata*-CO) and Roble (*Tabebuia rosea*-TR) representing a range of relative growth rates (Potvin and Gotelli, 2008). The 24 diversity plots of the plantation, measure approximately 45 × 45 m each, and consist of 6 six-species plots, 6 different three-species plots and 12 monocultures. The results presented in this paper encompass the first four years (2001–2005) of tree growth as well as an intensive campaign of soil and litter microfauna collection carried out in 2004.

Biomass was calculated for each individual tree from 2002 to 2005 using species specific allometric equations based on height and basal diameter (BD) for saplings <2 m (Coll et al., 2008). For trees (height >2 m), diameter at breast height (DBH) (1.3 m) was employed and allometric equations derived by Chave et al. (2005) were used. It should be noted that for the 2003 data, only BD was collected, even if a few trees more than 2 m in height were present. Species specific root to shoot ratios (Coll et al., 2008) were used to estimate below ground biomass thus allowing calculation of total biomass (above and below ground biomass). Individual tree data was then aggregated to obtain biomass estimates at the plot level. C stocks were estimated by converting the biomass data to carbon using species-specific percent C values collected in Panama in adjacent areas (Elias and Potvin, 2003).

During the two first growing seasons (July 2001 to December 2002), the intensity of damage inflicted to the seedlings for every episode of insect attacks was scored. During that period six severe attacks were observed, the only species showing no sign of massive insect attack was *Luehea seemanii*. For each episode, 60 seedlings were scored for each diversity level (10 seedlings per plot in the six-species plots, 20 in the three-species plots and 30 in monocultures) for a total of 1,260 seedlings. Sampling was regular, that is, the first seedling was scored when a regular number of seedlings were omitted to ensure an equal sample size for the three diversity levels (2 in six-species plots, 4 in three-species plots and 6 in monocultures). Attacks were scored as light, medium or severe based on the number of leaves eaten or of stems damaged. After the damages were scored either Furadan® insecticide-nematicide or Aribo (a pyrethroid-based insecticide) was applied to individual seedlings to prevent seedling death. The damages therefore did not necessarily lead to mortality. Nevertheless, the information allows us to assess the action of herbivores in our system. A weighted damage score was computed for each species, episode and diversity treatment as follow:

$$\text{Equation 1: } d_a = d_l + 2*d_m + 3*d_s + 4*d$$

Where d_a is the damage score, d_l the proportion of trees in each plot showing light damages, d_m medium damages, d_s severe damage and d the proportion of trees killed.

At the onset of the fourth rainy season, from May to late-August 2004, litter and soil arthropods were collected near 36 tree pairs including all six species growing in monoculture pairs and growing with the five other species (for example, COCO, CO-CA, CO-AE, CO-HC, CO-TR, CO-LS). Samples were collected beneath each focal tree and the neighboring tree was defined as the one in front of the sampling site at a 2.5 m distance. Litter arthropods were collected using a frame delineating 50 cm^2 underneath each main tree species. All the leaves and small woody litter (<2 mm in diameter) were gathered in plastic bags. Arthropods were extracted using Berlese funnels and a trapping device; samples were placed in BioQuip whirlbags containing 75% alcohol (Didham and Fagan, 2003). The samples were heated for 48 hours with a light bulb to promote migration into the trapping device. Litter arthropods were sorted and counted into major groups (that is, *Acari*) and orders. Soil arthropods were collected using a metal cylinder of 12 cm diameter by 6 cm depth, which consisted of two open ends and a handle. In the lab the soil was deposited in a Berlesse funnel and arthropods were extracted as described above. Finally, an extensive earthworm collection was performed: four replicates for every pair of tree species (36 tree combinations in total from the six tree species) to account for the densities (numbers m^{-2}) (Sarlo, 2006).

In addition, the biomass of each focal tree was estimated as explained above. Dry soil bulk density (SBD, g cm^{-3}) was measured for each sampling site, using a core of 3 cm in diameter and 6 cm in depth. After arthropods' extraction, the litter was dried in an oven (60°C) for three days and weighed. Finally, canopy projection of each sampled tree was estimated in the field by taking the distances in the north-south and west-east direction of the canopy. Canopy cover was then estimated based on the ellipse formula.

Data Analysis

The normality and skewness of the data were observed by plotting frequency distributions of the experimental variables. The *Cordia alliodora* plots were excluded from all biomass analyses because of their high rates of mortality (Potvin and Gotelli, 2008). To assess the pattern and shape of the response of tree biomass over time, repeated measured analyses of variances (ANOVARs) were performed. Each plot was subdivided in four subplots of 22 m × 20 m shown to be spatially independent from one another (Healy et al., 2008). The full data set (between 3,797 and 4,903 trees) was employed to calculate biomass for each of the remaining 88 subplots. In the analysis, Diversity, Plot (Diversity) and Subplot (Plot (Diversity)) were treated as between subject effects and year was within subject effects. The effect of species identity on individual tree biomass after four years of growth (2005) was further analyzed by using all individuals belonging to the 12 monocultures with 2 blocks for each monoculture (AE, CO, HC, LS and TR). Where results were significant, post hoc Tukey tests were performed.

The additive partitioning method proposed by Loreau and Hector (2001) was used to estimate the net biodiversity effect on plot basal area (BA) using 2006 data and to subsequently separate this into the corresponding complementarity and selection effects. The calculations were done (i) to compare mixture and monocultures plots (n = 12) and (ii) independently for the three- and six-species plots (n = 6). In the first case the BA of the mixture was compared with the average BA of all the best monocultures, while in the independent analysis of the three-species plots, the BA of that plot was only compared with the best BA of its constituent species in monocultures. In all cases, the biodiversity, complementarity and selection effects were tested by t-tests against a null value of zero, indicating no effect of biodiversity, complementarity or selection (Loreau and Hector, 2001).

Earthworm, litter and soil arthropod numbers and litter mass were converted to 1m^2 and transformed, log (x + 1), prior to analysis. Two indices of diversity—Shannon's and Simpson's—were computed to summarize information about the importance of the structure of the diversity of orders of Insecta and subclasses of *Arthropoda*. Shannon's index is estimated as:

$$\text{Equation 2: } H' = > p_i \ln p_i$$

Where p_i is the relative abundance of each order/subclass, calculated as the proportion of individuals of a given order/subclass to the total number of individuals in the community:

(n_i/N) with N being the total number of individuals and n is the number of individuals of a given order (or subclass). The formula for the Simpson's index is:

$$\text{Equation 3: } D => n_i(n_i - 1)/N(N - 1)$$

While Simpson's index is most sensitive to variation in the abundant taxa, Shannon's index captures changes in rare taxa. Simpson's index for an infinitely diverse community would take a value of 0 while, in contrast, Shannon's index increases with diversity. The variation in earthworm and arthropod data was examined at different scales relying on ANOVA for earthworms and MANOVA for arthropods. First two-way ANOVA or MANOVA with focal tree species and neighboring tree species as main effects were used. For earthworms only, significant differences among the pairs for each tree species were calculated using Newman-Keuls test ($p < 0.05$). A second analysis compared the effect of mixtures: for example, whether a sample was collected between monoculture vs. mixture tree pairs. The mixture effects were considered as a surrogate for the effect of diversity. ANCOVA was relied on to examine the importance of canopy cover, litter mass, tree biomass and dry soil bulk density taken independently as covariates for soil and litter insect numbers. Pearson's correlations were used to account for the variation in earthworm numbers due to canopy cover, litter mass, tree biomass, or dry SBD. All the statistical analyses were performed using SYSTAT 10.2 (Systat Software, 2002).

Finally, one way ANOVAs were used to test the effect of diversity on average attack scores per plot for individual species. The sum of D_a at the plot level was analyzed by Kruskal-Wallis for differences among diversity treatment.

RESULTS

Tree Growth, Diversity and Carbon Storage

Across treatments, the overall cell mean of tree biomass at the plot level was 0.079, 0.654, 1.165 and 2.195 t ha^{-1}, for year 1 through 4, with an overall increase in biomass of 0.686 t ha^{-1} yr^{-1} ($r^2 = 0.43$, $p < 0.000$). The greatest increase in biomass was between the third and fourth year of growth, with an average increase of 1.03 t ha^{-1} per plot. The ANOVAR used to analyze changes in subplot biomass through time unveiled a significant interaction ($F_{6,96} = 2.48$, $p = 0.028$) between year and diversity level. Biomass of the three-species plots is significantly higher than that of the other diversity levels and this effect increases through time (Figure 1). An in depth analysis was done with the 2005 data, corresponding to the fourth year of tree growth. An ANOVA of average tree biomass at the plot level in 2005, found a significant species effect ($F = 6.704$, $p < 0.05$). *A. excelsum* was found to exhibit the smallest average biomass while, *C. odorata* was the most productive species in the three- and six-species mixtures. Species contribution to plot biomass varies across species. In both the three- and six-species plots, *C. alliodora* contributes less biomass than expected based on the number of individuals planted while *C. odorata* contributes more (Figure 2).

The additive partitioning method (Loreau and Hector, 2001) was used to compare the basal area of mixtures with monocultures. This analysis unveils a significant positive effect of complementarity ($t = 2.64$, $p = 0.023$) and a significant negative effect of selection ($t = -2.21$,

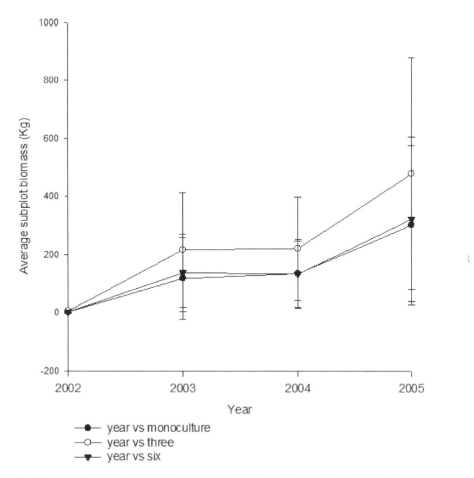

FIGURE 1. Changes in average Subplot biomass through time between planting (2001) and the fourth growing season (2005). Data are for monocultures, three- and six-species subplots (n = 24 for three- and six-species plots and n = 40 for monocultures).

p = 0.049). These effects being of the same magnitude, no significant net biodiversity effect was detected (Figure 3). The independent analysis of three-species plots unveils a significant net biodiversity effect (t = 2.6, p = 0.048) with the complementarity effect being 42% higher than the selection effect. For six-species plots, neither selection nor complementarity was significantly different from the null value of 0.

The biomass data at the plot level were used to estimate C storage after four years. The highest value was 2.76 tC ha^{-1} after four years of growth for a three-species plot planted with *C. odorata, H. crepitans* and *L. seemannii* (Figure 4). The lowest carbon storage value was that of the *A. excelsum* monocultures with 0.21 tC ha^{-1}. While the three-species plots store on average more carbon than the six-species plots or the monocultures (respectively 1.511 + 0.865 tC ha^{-1}, 0.963 + 0.732 tC ha^{-1} and 0.951 + 0.630 tC ha^{-1}), the difference among the diversity treatment is not statistically significant. This is apparently related to the high variability observed among plots within a diversity level (Figure 4).

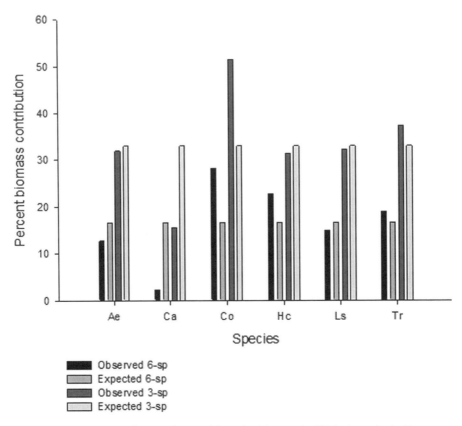

FIGURE 2. Species-specific contribution (%) to plot biomass in 2005 when planted in monocultures or in mixtures (three- and six-species plots).

Soil Fauna Diversity

A total of 20,579 soil organisms belonging to 18 different orders or groups were collected in the soil samples and another 66,563 organisms belonging to 21 different orders or groups were collected in the litter samples. For all the tree species in the plantation, the most abundant *Insecta* order was *Collembola* which accounted for more than 60% of the total numbers of organisms sampled under CO, LS and CA (Image 14, see photospread). None of the statistical analysis unveiled significant variation in the abundance, the diversity or composition of soil arthropods.

Conversely, the effect of mixture (monoculture vs. mixed species pairs) on the number of litter arthropods was statistically significant (Pillai Trace value = 0.130, p = 0.040), with fewer arthropods found below trees growing in monocultures (2,266 + 2,158 ind m^{-2}) than in mixed species (2,548 + 4,275 ind m^{-2}) pairs. Mites contributed more to the total arthropod numbers in monoculture than in mixed pairs, whereas springtails were more abundant in mixed pairs than in monocultures (Image 15, see photospread). *Homopterans* were especially sensitive to mixture vs. monoculture plantings (F$_{1, 142}$ = 7.219, p = 0.008) and were found at a higher density in monocultures, 16.5 ind m^{-2}, than in mixed plots, 5.5 ind m^{-2}. This is

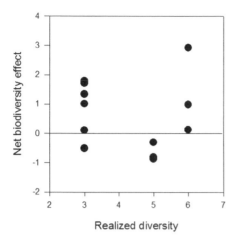

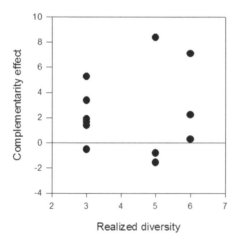

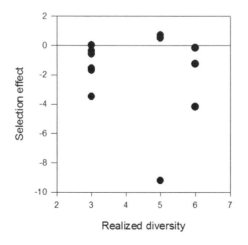

FIGURE 3. Additive partitioning of the net biodiversity effect on basal area (2006) at the plot level (m² ha⁻¹) into complementarity and selection effects. All mixture plots were compared with the six best monocultures. The null hypothesis of no effect is represented by the zero line. Because *Cordia alliodora* died in three of the six-species plots, we report realized diversity (3, 5 and 6) rather than original planting diversity (3 and 6).

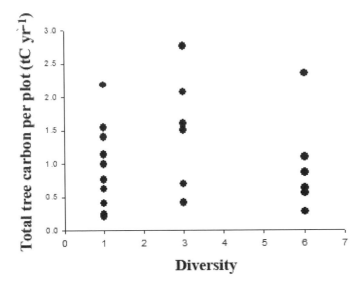

FIGURE 4. Carbon stocks per plot for each the 12 monocultures, the 6 three-species and the six-species plots. Species identity are reported in the figure as 0 (CA), 1 (AE), 2 (LS), 3 (HC), 4 (TR) and 5 (CO). For species codes refer to Table 2.

especially true for *C. odorata* which sustained 35.3 ± 1.8 ind m^{-2} in monoculture pairs but only 2.9 ± 1.4 ind m^{-2} in mixed pairs.

ANCOVA unveiled a constantly significant mixture effect for Shannon's and Simpson's indices of diversity with various covariates (Table 1). On average, monocultures had a higher Shannon's index (0.717 ± 0.048) than mixtures (0.563 ± 0.030) and mixtures had a higher Simpson's index (0.436 ± 0.032) than did monoculture plots (0.270 ± 0.040) indicating that mixed-species pairs sustain less diverse arthropod communities than monoculture pairs (Table 1). The maximum Simpson's value, 0.809, was found for mixed pairs of *C. odorata* and *C. alliodora* where *Collembolans* alone accounted for 89% of the species present. On the other hand, the minimum Simpson's value, 0.160, was found for monoculture plots of *L. seemanii*. Of the four covariates that were used in the analyses only soil bulk density had an apparent significant effect on Simpson's index (Table 1).

MANOVA unveiled a significant effect of tree species (Pillai Trace value = 0.647, p = 0.014) on the number of litter arthropods. The highest number of arthropods was found underneath *L. seemanii* and *A. excelsum* whereas *T. rosea* contained the lowest number for total arthropod counts (Image 16A, photospread). Two taxa, namely *Araneae* and *Coleoptera*, drove the significant response of tree species identity (respectively $F_{5,108}$ = 5.317, p = 0.0001; $F_{5,108}$ = 2.324, p = 0.048). Spiders ranged from 3.91 ± 1.32 ind m^{-2} in *T. rosea* to 19.18 ± 5.13 ind m^{-2} in *L. seemanii* a significant difference among tree species while beetles ranged between 25.18 ± 1.39 ind m^{-2} for *C. odorata* and 78.80 ± 1.21 ind m^{-2} for *A. excelsum* (Image 16B, photospread). The effect of tree species identity on Shannon's or Simpson's diversity indices was nonsignificant. This was also the case for the other measures of diversity considered in the study (total number of individuals of insects, total number of *Insecta* orders and total number of arthropod subclasses).

While the mixture effect was not statistically significant for the number of earthworms, significant differences in earthworm numbers among different tree species pairs were unveiled by Newman-Keuls test (p < 0.05). Earthworm density was higher in monoculture pairs of

TABLE 1. Results for ANCOVA for the mixture effect on diversity (Shannon and Simpson indices) and with canopy cover, litter mass, tree biomass and dry soil bulk density as covariates. Statistically significant differences are shown in bold.

Shannon's index	Mixture	$F_{1,33} = 4.375$	**p = 0.044**
	Canopy cover	$F_{1,33} = 1.666$	p = 0.206
	Mixture	$F_{1,33} = 4.623$	**p = 0.039**
	Soil bulk density	$F_{1,33} = 0.013$	p = 0.909
	Mixture	$F_{1,33} = 4.742$	**p = 0.037**
	Tree biomass	$F_{1,33} = 3.456$	p = 0.072
Simpson's index	Mixture	$F_{1,33} = 4.581$	**p= 0.040**
	Canopy cover	$F_{1,33} = 2.599$	p= 0.116
	Mixture	$F_{1,33} = 5.114$	**p = 0.030**
	Soil bulk density	$F_{1,33} = 5.275$	**p = 0.028**
	Mixture	$F_{1,33} = 4.850$	**p = 0.035**
	Tree biomass	$F_{1,33} = 0.093$	p = 0.762
	Mixture	$F_{1,33} = 4.533$	**p = 0.041**
	Litter mass	$F_{1,33} = 0.091$	p = 0.765

TR and LS but lower for the other four species (Table 2). Furthermore, the tree species main effect was statistically significant ($F_{5,91} = 3.28$, p = 0.009) and explained 28% of the variation in earthworm density. Fisher's a posteriori test showed that the number of earthworms below *H. crepitans* (339.40 ind m^{-2}) was significantly greater (p < 0.05) than below other tree species: *A. excelsum* (103.9 ind m^{-2}), *L. seemanii* (99.69 ind m^{-2}), *T. rosea* (72.79 ind m^{-2}), *C. alliodora* (57.88 ind m^{-2}) and *C. odorata* (47.98 ind m^{-2}).

The effect of neighboring tree identity on earthworm numbers was not significant, however the interaction of species and neighbor on earthworm numbers was significant ($F_{25,91} = 2.06$, p = 0.007). Highly variable earthworm densities were observed for the same focal species as neighbors changed (Table 2). For all the possible combinations, the greatest earthworm numbers were found near *H. crepitans* when paired with *C. odorata*. When T. rosea or *L. seemanii* was grown in monoculture pairs the density of earthworms in the soil are greater than when these two species were grown in mixed pairs. The lowest earthworms' density (1.57 ind m^{-2}) was associated with the combination of *C. odorata* (main species) and *A. excelsum* (neighboring species). In addition, canopy cover, dry litter mass, tree biomass and dry soil bulk density were measured and analyzed for their possible effects on earthworm. Simple Pearson's correlations were performed between earthworm numbers with canopy cover, litter mass, tree biomass and dry soil bulk density. Earthworm numbers and canopy cover were significantly correlated, $r^2 = 0.204$ ($\chi^2 = 5.275$, p = 0.022).

Insect damages to seedlings and saplings in the first two years of the plantation were used as an initial measure of the relationship between tree growth and insect diversity. The

TABLE 2. Mean earthworm densities for all species combinations. Tree species code: AE = *Anacardium excelsum*, CA = *Cordia alliodora*, CO = *Cedrela odorata*, HC = *Hura crepitans*, LS = *Luheia seemanii*, and TR = *Tabebuia rosea*. For the upper two species in the table, bold indicates earthworm density values for monocultures (maximum value). For the lower two species, italics indicates minimum values for the pair CO-AE and AE-CO. Significant differences (Newman-Keuls test, $p < 0.05$) for the pairs among each tree species are expressed by different letters set as superscripts.

Species	Neighbor	Earthworm Density (ind/m²)	Species	Neighbor	Earthworm Density (ind/m²)
TR	TR	**377.4** [c]	LS	LS	**305.9** [e]
	CO	94.72 [d]		CO	228.6 [d]
	CA	249.0 [c]		CA	42.65 [a, b]
	HC	270.6 [b]		HC	54.21 [b]
	AE	16.58 [a]		AE	134.8 [c]
	LS	7.13 [a]		TR	28.44 [a]
CO	CO	220.3 [a]	AE	AE	171.6 [a]
	CA	11.16 [b]		CO	*7.610* [b]
	HC	348.1 [c]		CA	31.43 [c]
	AE	*1.570* [d]		HC	209.4 [d]
	TR	282.8 [e]		TR	178.1 [a]
	LS	28.85 [f]		LS	333.2 [e]
HC	HC[c]	479.84 [e]	CA	CA	10.97 [b]
	CO[d]	752.36 [d]		CO	87.31 [a]
	CA[c]	563.93 [c]		HC	76.45 [a]
	AE[a]	286.74 [a]		AE	45.88 [c]
	TR[b]	93.41 [b]		TR	147.59 [d]
	LS[a]	321.11 [a]		LS	114.35 [e]

average insect attack scores were similar across diversity levels with 1.16 + 0.56 for monocultures, 1.17 + 0.68 for three-species plots and 1.14 + 0.56 for the six-species plots and the diversity effect is not significant ($F_{2,63} = 0.015$, $p = 0.985$). However, the sum of insect attacks score was significantly higher in six-species plots (6.83 + 0.23) than in either the three-species plots (3.25 + 1.51) or the monocultures (2.78 + 1.61) ($F_{2,14} = 17.95$, $p < 0.001$). Pearson's correlation coefficient did not detect any significant relationship between the summed attack scores and plot biomass (at the end of 2002) ($\chi^2 = 0.036$, $p = 0.850$). Two species are apparently most sensitive to diversity levels; average attack scores of both *C. odorata* and *A. excelsum* were significantly less in mixture than in monoculture plots ($F_{2,8} = 5.69$, $p < 0.05$ and $F_{2,8} = 4.52$, $p < 0.05$ respectively). Insect damages were 40% higher on trees of *C. odorata* growing in monoculture than on those growing in six-species plots (Table 3) while, for *A. excelsum*, attacks were 25% lower in three-species plots than when the species was growing alone.

TABLE 3. Damage scores values are mean ± standard deviation. Differences between the treatments were analyzed by ANOVA and means followed by the same letter are not significantly different.

Damage Score	6-Species Plots	3-Species Plots	Monocultures
C. alliodora	0.442 + 0.332	0.192 + 0.063	0.183 + 0.071
L. seemanii	0	0	0
H. crepitans	1.275 + 0.308	1.417 + 0.419	1.150 + 0.257
A. excelsum	1.08 + 0.13	1.0 + 0.0	1.33 + 0.19
C. odorata	0.733 + 0.175	0.786 + 0.208	1.217 + 0.118
T. rosea	2.017 + 0.279	2.183 + 0.202	1.917 + 0.212

DISCUSSION

Trees' Biodiversity Effects

Young reforestation plots in Brazil and Costa Rica were found to exhibit biomass increments ranging between 1.6–29.8 t ha^{-1} yr^{-1}, with differences resulting from species identity and the age of the plantation (Fonseca et al., this volume; Lugo et al., 1988; Stanley and Montagnini, 1999). On a per hectare basis, our values of total tree biomass lie at the smaller end of the spectrum (four years: 2.195 t ha^{-1}). Early experimental studies with trees reported that mixtures are often more productive than monocultures (Bauhus et al., 2000; Erskine et al., 2006; Forrester et al., 2004; Khanna, 1997; Menalled et al., 1998; Parrota et al., 1997, 1999; Sayyad et al., 2006). Species mixtures can out produce their corresponding monocultures via a number of different mechanisms; for example intraspecific competition could be minimized, thus enabling a more efficient use of nutrients (Montagnini et al., 1995). This may occur through niche differentiation; whereby the different soil strata are occupied differentially by the root systems of the different species (Sayyad et al., 2006) and or via canopy stratification which enables the more efficient use of solar energy (Menalled et al., 1998).

In Sardinilla, biodiversity apparently enhances individual tree growth, without affecting tree mortality, resulting in increased productivity at the plot level (Potvin and Gotelli, 2008). The additive partitioning method suggests that complementarity, especially in three-species plots, plays a major role in the system. Such complementarity results in higher biomass with consequences for C storage. With ~3 tC ha^{-1}, the plot that stored the most C was a three-species mixture containing *Cedrela odorata, Hura crepitans* and *Luhea seemanii*. Another three-species plot (*Tabebuia rosea, Anacardium excelsum* and *Luehea seemanii*) and one of the six-species plots also stored more than 2 tC ha^{-1}. Carbon models developed for Central America estimated a C storage potential of 6 tC ha^{-1} after three years with a stem density of 988 individuals (Keogh, 2004). In Sardinilla, average stem density of the three-species plot was only 704 + 194 after four years of growth. The C storage capacity of these plots is thus slightly lower than that estimated for commercial teak plantation of a similar age, in part because of the lower tree density.

Several studies have demonstrated that trees acquire, store and recycle nutrients in different ways (Cuevas and Lugo, 1998; Forrester et al., 2005; Shiels, 2006). The influence of a species on ecological processes will depend on the traits of this individual species (Tilman, 1999).

In our study, differential species contribution to subplot biomass suggests that species choice will also play a central role in determining C storage capacity. Across the plantation, the monocultures of *C. odorata* showed superior C storage than any other monocultures and this species contributed more than expected to biomass of mixtures, be they three- or six-species plots. In contrast, two of the three-species plots and the monocultures established with AE stored the least C. In combination, the results from the additive partitioning analysis, the biomass and C storage demonstrate that species identity is an important component of the biodiversity effect. Species specific traits will govern how species will interact with each other.

Soil Fauna Response to Tree Diversity

Several studies in grassland and agroecosystems provide evidence for the effect of plant species on the community of arthropods, especially insects (Haddad et al., 2001; Knops et al., 1999; Siemann et al., 1998). Some show that decreasing plant diversity has a negative effect on soil biota (Niklaus et al., 2001) whereas others found no consistent responses (Wardle and Lavelle, 1997). Not so well studied is the effect of diversity of native tree species on the arthropod diversity in tropical plantation forests, as considered here. This study's hypothesis was that the tree species mixture and identity would affect the diversity of litter and soil arthropods as well as the number of earthworms. Contrary to this hypothesis, neither species or a mixture was found to have an effect on soil arthopods, probably because the soil is not as sensitive as the litter such that effects of tree species on soil fauna might only be revealed in the long term. Similar results were found on other soil organisms, microbes and nematodes in a grassland study (Gastine et al., 2003).

While the mixture effect was not significant for earthworms and soil arthropods, mixture pairs sustained more litter arthropod than the monocultures. Knops et al. (1999) also reported that more diverse systems promoted greater diversity at other trophic levels. Greater availability of plant resources or vegetation structure might account for this (Haddad et al., 2001).

Homoptera is an interesting taxon; contrary to the general trend, this study found that they were more abundant in monocultures than on mixture pairs. A possible explanation is that *Homoptera* is more host specific and more sedentary than most herbivores (Strong et al., 1984). For example, the family *Cicadellidae* has been shown to decrease in abundance as a response to an increase in plant species number probably because it is a specialized family and reproduces sexually (Koricheva et al., 2000). As plant-species richness increases, the density of any given species potentially decreases at the landscape level. For an insect herbivore with host specific feeding habits, the indirect effect of higher plant diversity is therefore a reduction in food resources (Joshi et al., 2004; Koricheva et al., 2000). In Sardinilla, the sparse availability of host plants could explain the low populations of *Homoptera* in mixture pairs. However, not all *Homopterans* show the same response to increases in plant diversity (Koricheva et al., 2000). Watt (1997) found *Homopterans* to be more abundant in disturbed plots as compared to uncleared forest plots in Cameroon.

As expected, MANOVA, analyzing the effect of tree species identity on the entire litter arthropod fauna, unveiled that tree species exert a significant effect on the composition of this fauna. The effect was driven by beetles and spiders. The highest abundances for *Acari* and *Collembola* among all groups considered is coherent with other reports (Badejo and Tian,

1999; Heneghan et al., 1998). The two litter arthropod taxa that drove the response of tree species identity, spiders and beetles, showed highest abundances on *Luheea seemanii* and *Anacardium excelsum*, respectively. Furthermore, of all the tree species, *Hura crepitans*, sustained the greatest numbers of earthworms. Among the mechanisms that can explain the responses of soil and litter biota to the identity of tree species are the quantity and quality of resources produced, competitiveness against microorganisms for nutrients and formation of habitats (Wardle, 2002). For instance, trees containing high nutrient concentrations and low level of phenolic compounds can support a greater number of invertebrate species. Examples of direct effects of trees on invertebrates include *Coleoptera* feeding on high nutrient tree leaves, while indirect effects include spiders feeding on herbivorous beetles.

Similar densities of earthworms to those reported here were observed by Barros et al. (2003) for agroforestry sites in Amazonia, with densities ranging from 107 ind m^{-2} to 323 ind m^{-2} in the first 25 cm of the soil. Recovering pastures to young recovered forests in Puerto Rico also contained earthworm densities in the range of this study, from 273.7 ind m^{-2} to 172.4 ind m^{-2} in the top 25 cm of the soil (Leon et al., 2003). Undisturbed forests appear to contain fewer numbers of earthworms than human impacted ecosystems. Trees have been shown to be important in maintaining native earthworms (Fragoso et al., 1997) indicating that some species of trees are more beneficial for earthworms than others. According to several studies, earthworm density decreases as successional stages increase (Fragoso and Lavelle, 1992; Leon et al., 2003; Zou and Gonzalez, 1997).

Ecosystem Linkages: Tree Diversity, Litter Decomposition, Soil Respiration and Faunal Responses

It has been hypothesized that plant species growing in diverse communities gain protection against herbivore attack (Carson et al., 2004) in comparison to one species stands. A review using a meta-data analysis to examine a data set of 54 independent studies compared herbivore damages of single vs. mixed-species plots (Jactel et al., 2005). The authors report that mixed-stands suffer less pest damage than monocultures and proposed three hypotheses to explain the mechanisms underlying tree species diversity stands and their resistance to pest insect infestations: host accessibility to pests is decreased, the impact of natural enemies is more effective, and pest shift among host tree species. Results from this study on the damage to seedlings from herbivorous insects are coherent with this hypothesis since *C. odorata* trees grown in monoculture plots suffered significantly more attacks than when in mixture plots. From an applied perspective the result is important because *C. odorata* is a prime native timber in Panama.

Multiple studies are carried out in the Sardinilla plantation allowing to relate different ecosystem processes. These results seem to suggest an important link between abundance of earthworms and litter decomposition. Scherer-Lorenzen et al. (2007) reported that *H. crepitans* had significantly higher decaying rates for its litter than the other tree species in the plantation, possibly due to a lower lignin concentration. These results go hand in hand with the present study: *H. crepitans*, yielded significantly greater number of earthworms than the rest of the tree species, and when the interaction of focal tree species and neighbor was considered, *H. crepitans* also contained the highest numbers of earthworms. This raises the possibility of a complex feedback between tree identity, litter decomposition and soil fauna.

This study purposes that the rapid decomposition of *H. crepitans* leaves is due to its low lignin content and that the low lignin content makes it good quality litter for earthworms. As a consequence, high abundance of earthworms under *H. crepitans* breaks down the litter at faster rates than under the other tree species. Additionally, Scherer-Lorenzen et al. (2007) found that *A. excelsum* was the most recalcitrant tree species and, in the present study, when *A. excelsum* is the neighboring species, the number of earthworms under the focal tree falls in the lower ranges compared with the other neighbors.

Moreover, Murphy et al. (2008) found that, in Sardinilla, soil respiration of monoculture pairs of *H. crepitans*, *L. seemanni* and *T. rosea* during the dry season was higher than that of pairs composed of HC-LS, HC-TR, LS-TR. This study has shown that earthworm numbers were higher in these monoculture plots than in the above-mentioned combinations. Earthworms contribute to soil respiration by providing high quality carbon for microbial processes (Li et al., 2002). Therefore, earthworm's contribution to soil respirations rates might be significant. Murphy et al. (2008) also reported lower soil respiration rates in three-species plots than in monocultures and explained this effect by increased crown cover in these plots. Interestingly, the numbers of earthworm were significantly correlated with canopy cover. Most probably, earthworms gather in areas where better microclimatic conditions are generated by the shade of the trees. This study proposes that increased crown cover affects soil moisture and temperature with direct effect on soil respiration and on earthworm density, which in turn would affect soil respiration. The Sardinilla results therefore emphasize the complex direct and indirect effects of trees on soil fauna with feedback on ecosystem processes (Neher, 1999).

The context of the current carbon market is favorable to tropical plantations (Olschewski et al., 2005), however concern is rising over the biodiversity effect of large-scale monoculture plantations (Cyranoski, 2007). Although initial tree C storage might be somewhat lower than with teak, native species plantations can be a good mitigation strategy in the context of climate change, in that they entail significant biodiversity benefits with correlated impact on ecosystem C cycling. Results from this study further emphasize that, by supporting differing groups of soil fauna, the individual species of trees act as building blocks for diversity. Therefore the design of mixed-species plantation with native species should pay careful attention to species choice as an important determinant of potential C stock and biodiversity.

ACKNOWLEDGMENTS

Catherine Potvin would like to acknowledge the constant support of the Smithsonian Tropical Research Institute and of a Discovery Grant from the Natural Science and Engineering Research Council of Canada. Chrystal Healy was supported by a Québec's FQRNT graduate fellowship. Malena Sarlo benefited from support from the International Development Research Center of Canada. We are indebted to Jose Monteza and Anayansi Valderama for their excellent technical support.

REFERENCES

Amoroso, M. M., and E. C. Turnblom. 2006. Comparing Productivity of Pure and Mixed Douglas Fir and Western Hemlock Plantation in the Pacific Northwest. Ottawa, ON, Canada: NRC Research Press.

Anderson, J. M. 1988. Invertebrate-Mediated Transport Processes in Soils. *Agriculture, Ecosystems and Environment*, 24:5–19.

Anderson, J. M., and J. S. I. Ingram. 1993. *Tropical Soil Biology and Fertility: A Handbook of Methods*. Wallingford, UK: CAB International.

Badejo, M. A., and G. Tian. 1999. Abundance of Soil Mites under Four Agroforestry Tree Species with Contrasting Litter Quality. *Biology and Fertility of Soils*, 30:107–112.

Barros, E., A. Neves, E. Blanchart, E. C. M. Fernandes, E. Wandelli, and P. Lavelle. 2003. Development of the Soil Macrofauna Community under Silvopastoral and Agrosilvicultural Systems in Amazonia. *Pedobiologia*, 47:273–280.

Bauhus, J., P. K. Khanna, and N. Menden. 2000. Aboveground and Belowground Interactions in Mixed Plantations of *Eucalyptus globules* and *Acacia mearnsii*. *Canadian Journal of Forestry Research,* 30:1886–1894.

Bristow, M., J. D. Nichols, and J. K. Vanclay. 2006. Growth and Species Interactions of *Eucalyptus pellita* in a Mixed and Monoculture Plantation in the Humid Tropics of North Queensland. *Forest Ecology and Management*, 233:193–194.

Cardinale, J. 2006. Effects of Biodiversity on the Functioning of Trophic Groups and Ecosystems. *Nature*, 443:989–992.

Carnus, J., J. Parrotta, E. Brockerhoff, M. Arbez, H. Jactel, A. Kremer, D. Lamb, K. O'Hara, and B. Walters. 2006. Planted Forests and Biodiversity. *Journal of Forestry,* 104:65–77.

Carson, W. P., J. Patrick Cronin, and Z. T. Long (eds.). 2004. *A General Rule for Predicting When Insects Will Have Strong Top-Down Effects on Plant Communities: On the Relationship between Insect Outbreaks and Host Concentration*. Ecological Studies. Berlin-Heidelberg: Springer-Verlag.

Chave, J., C. Andalo, S. Brown, M. A. Cairns, J. Q. Chambers, D. Eamus, H. Folster, F. Fromard, N. Higuchi, T. Kira, J. P. Lescure, B. W. Nelson, H. Ogawa, H. Puig, B. Reira, and T. Yamakura. 2005. Tree Allometry and Improved Estimation of Carbon Stocks and Balance in Tropical Forests. *Oecologia,* 145:87–99.

Coll, L., C. Potvin, C. Messier, and S. Delagrange. 2008. Root Architecture and Allocation of Eight Tropical Species with Different Shade Tolerance in Open-Grown Mixed Plantations in Panama. *Trees—Structure and Function,* 22:585–596.

Cuevas, E., and A. E. Lugo. 1998. Dynamics of Organic Mater and Nutrient Return from Litterfall in Stands of Ten Tropical Tree Plantation Species. *Forest Ecology and Management,* 112:263–279.

Cyranoski, D. 2007. Logging: The New Conservation. *Nature,* 455:608–610.

Didham, R. K., and L. L. Fagan. 2003. Project IBISCA—Investigating the Biodiversity of Soil and Canopy Arthropods. *Weta,* 26:1–6.

Elias, M., and C. Potvin. 2003. Assessing Intra- and Inter-Specific Variation in Trunk Carbon Concentration for 32 Neotropical Tree Species. *Canadian Journal of Forest Research,* 33:1039–1045.

Erskine, P. D., D. Lamb, and M. Bristow. 2006. Tree Species Diversity and Ecosystem Function: Can Tropical Mixed-Species Plantations Generate Greater Productivity? *Forest Ecology and Management,* 233:205–210.

Erwin, T. L. 1982. Tropical Forests: Their Richness in Coleoptera and Other Arthropod Species. *Coleopterist Bulletin,* 36:74–75.

Forrester, D. I., H. Bauhus, and A. L. Cowie. 2005. On the Success and Failure of Mixed Species Tree Plantations: Lessons Learned from a Model System of Eucalyptus Globules and *Acacia mearnsii*. *Forest Ecology and Management,* 209:147–155.

Forrester, D. I., J. Bauhus, and P. K. Khanna. 2004. Growth Dynamics in a Mixed-Species Plantation of Eucalyptus Globules and *Acacia mearnsii*. *Forest Ecology and Management,* 193:81–95.

Fragoso, C., G. G. Brown, J. C. Patron, E. Blanchart, P. Lavelle, B. Pashanasi, B. Senapati, and T. Kumar. 1997. Agricultural Intensification, Soil Biodiversity and Agroecosystem Function in the Tropics: The Role of Earthworms. *Applied Soil Ecology,* 6:17–35.

Fragoso, C., and P. Lavelle. 1992. Earthworm Communities of Tropical Rain Forests. *Soil Biology and Biochemistry,* 24:1397–1408.

Fridley, J. D. 2003. Diversity Effects on Productivity in Different Light and Fertility Environments: An Experiment with Communities of Annual Plants. *Journal of Ecology,* 91:396–406.

Gastine, A., M. Scherer-Lorenzen, and P. W. Leadley. 2003. No Consistent Effects of Plant Diversity on Root Biomass, Soil Biota and Soil Abiotic Conditions in Temperate Grassland Communities. *Applied Soil Ecology,* 24:101–111.

Grant, J. C. 2006. Five Year Results from a Mixed-Species Spacing Trial with Six Subtropical Rainforest Tree Species. *Forest Ecology and Management,* 233:309–314.

Haddad, N. M., D. Tilman, J. Haarstad, M. E. Ritchie, and J. M. H. Knops. 2001. Contrasting Effects of Plant Richness and Composition on Insect Communities: A Field Experiment. *American Naturalist,* 158:17–35.

Healy, C., N. J. Gotelli, and C. Potvin. 2008. Partitioning the Importance of Diversity and Environmental Heterogeneity in a Tropical Tree Plantation. *Journal of Ecology,* 96:903–913.

Heneghan, L., D. C. Coleman, X. Zou, D. A. Crossley Jr., and B. L. Haines. 1998. Soil Microarthropod Community Structure and Litter Decomposition Dynamics: A Study of Tropical and Temperate Sites. *Applied Soil Ecology,* 9:33–38.

Hooper, D. U., F. S. Chapin, J. J. Ewel, A. Hector, P. Inchausti, S. Lavorel, J. H. Lawton, D. M. Lodge, M. Loreau, S. Naeem, B. Schmid, H. Setala, A. J. Symstad, J. Vandermeer, and D. A. Wardle. 2005. Effects of Biodiversity on Ecosystem Functioning: A Consensus of Current Knowledge. *Ecological Monographs,* 75:3–35.

Jactel, H., E. Brockerhoff, and P. Duelli. 2005. A Test of the Biodiversity-Stability Theory: Meta-Analysis of Tree Species Diversity Effects on Insect Pest Infestations, and Reexamination of Responsible Factors. In *Forest Diversity and Function: Temperate and Boreal System,* Ecological Studies 176, ed. M. Scherer-Lorenzen, C. Körner, and E.-D. Schulze, pp. 235–261. Berlin Heidelberg: Springer-Verlag.

Jones, C. G., J. H. Lawton, and M. Shachak. 1994. Organisms as Ecosystem Engineers. *Oikos,* 79:373–386.

Joshi, J., S. J. Otway, J. Koricheva, A. B. Pfisterer, J. Alphei, B. A. Roy, M. Scherer-Lorenzen, B. Schmid, E. Spehn, and A. Hector. 2004. Bottom-Up Effects and Feedbacks in Simple and Diverse Experimental Grasslands Communities. In *Insects and Ecosystem Function,* ed. W. W. Weisser and E. Siemann, pp. 413. Berlin Heidelberg: Springer-Verlag.

Keogh, R. M. 2004. Carbon Models and Tables for Teak (*Tectona grandis* Linn f.) Central America and Carribean. Dublin: Coillte Consult, International Teak Unit.

Khanna, P. K. 1997. Comparison of Growth and Nutrition of Young Monocultures and Mixed Stands of Eucalyptus Globules and *Acacia mearnsii. Forest Ecology and Management,* 94:105–113.

Knops, J. M. H., D. Tilman, N. M. Haddad, S. Naeem, C. E. Mitchell, J. Haarstad, M. E. Ritchie, K. M. Howe, P. B. Reich, E. Siemann, and J. Groth. 1999. Effects of Plant Species Richness on Invasion Dynamics, Disease Outbreaks, Insect Abundances and Diversity. *Ecology Letters,* 2:286–293.

Koricheva, J., C. P. H. Mulder, B. Schmid, J. Joshi, and K. Huss-Danell. 2000. Numerical Responses of Different Trophic Groups of Invertebrates to Manipulations of Plant Diversity in Grasslands. *Oecologia,* 125:271–282.

Lal, R. 1987. Earthworms. In *Tropical Ecology and Physical Edaphology,* ed. R. Lal, pp. 732. Chinchester, West Sussex, UK: Wiley Interscience.

Lanta, V., and J. Leps. 2006. Effects of Functional Group Richness and Species Richness in Manipulated Productivity-Diversity Studies: A Glasshouse Pot Experiment. *Acta Oecologica,* 29:85–96.

Leon, Y. S.-D., X. Zou, S. Borges, and H. Ruan. 2003. Recovery of Native Earthworms in Abandoned Tropical Pastures. *Conservation Biology,* 17:999–1006.

Li, Xuyong, M. C. Fisk, T. J. Fahey, and P. J. Bohlen. 2002. Influence of Earthworm Invasion on Soil Microbial Biomass and Activity in a Northern Hardwood Forest. *Soil Biology and Biochemistry,* 34: 1929–1937.

Loreau, M., and A. Hector. 2001. Partitioning Selection and Complementarity in Biodiversity Experiments. *Nature,* 412:72–76.

Lugo, A. E., S. Brown, J. and Chapman. 1988. An Analytical Review of Production Rates and Stemwood Biomass of Tropical Forest Plantations. *Forest Ecology and Management* 23:179–200.

Menalled, F. D., M. J. Kelty, and J. J. Ewel. 1998. Canopy Development in Tropical Tree Plantations: A Comparison of Species Mixtures and Monocultures. *Forest Ecology and Management,* 104:249–263.

Montagnini, F., A. Fanzeres, and S. G. DaVinha. 1995. The Potentials of 20 Indigenous Tree Species for Soil Rehabilitation in the Atlantic Forest Region of Bahia, Brazil. *Journal of Applied Ecology,* 32:841–856.

Montagnini, F., and C. Porras. 1998. Evaluating the Role of Plantation as Carbon Sinks: An Example of an Integrative Approach from the Humid Tropics. *Environmental Management,* 22:459–470.

Murphy, M., T. Balser, N. Buchmann, V. Hann, and C. Potvin. 2008. Linking Tree Biodiversity to Belowground Process in a Young Tropical Plantation: Impacts on Soil CO_2 Flux. *Forest Ecology and Management,* 255:2577–2588.

Naeem, S., and J. P. Wright. 2003. Disentangling Biodiversity Effects on Ecosystem Functioning: Deriving Solutions to a Seemingly Insurmountable Problem. *Ecology Letters,* 6:567–579.

Neher, D. A. 1999. Soil Community Composition and Ecosystem Processes: Comparing Agricultural Ecosystems with Natural Ecosystems. *Agroforestry Systems,* 45:159–185.

Niklaus, P. A., E. Kandeler, P. W. Leadley, B. Schmid, D. Tscherko, and C. Korner. 2001. A Link between Plant Diversity, Elevated CO_2 and Soil Nitrate. *Oecologia,* 127:540–548.

Novotny, V., Y. Basset, S. E. Miller, G. D. Weiblen, B. Bremer, L. Cizek, and P. Drodz. 2002. Low Host Specificity of Herbivorous Insects in a Tropical Forest. *Nature,* 416:841–844.

Olschewski, R., P. C. Benítez, P. C., G. H. J. de Koning, and T. Schlichter. 2005. How Attractive Are Forest Carbon Sinks? Economic Insights into Supply and Demand of Certified Emission Reductions. *Journal of Forest Economics,* 11:77–94.

Parrotta, J. A. 1999. Productivity, Nutrient Cycling, and Succession in a Single- and Mixed-Species Plantations of *Casuarina equisetifolia, Eucalyptus robusta,* and *Leucaena leucocephala* in Puerto Rico. *Forest Ecology and Management,* 124:45–77.

Parrota, J., J. Turnbull, and N. Jones. 1997. Catalyzing Native Forest Regeneration on Degraded Tropical Lands. *Forest Ecology and Management,* 99:1–7.

Potvin, C., and N. J. Gotelli. 2008. Biodiversity Enhances Individual Performance but Does Not Affect Survivorship in Tropical Trees. *Ecology Letters,* 11:217–223.

Potvin, C., E. Whidden, and T. Moore. 2004. A Case Study of Carbon Pools under Three Different Land Uses in Panama. *Climatic Change,* 67:291–307.

Redondo-Brenes, A., and F. Montagnini. 2006. Growth, Productivity, Aboveground Biomass, and Carbon Sequestration of Pure and Mixed Native Tree Plantations in the Caribbean Lowlands of Costa Rica. *Forest Ecology and Management,* 232:168–178.

Sarlo, M. 2006. Individual Tree Species Effects on Earthworm Biomass in a Tropical Plantation in Panama. *Caribbean Journal of Science,* 42:419–427.

Sayyad, E., M. H. Seyed, M. Jamshid, M. Reza, G. J. Seyed, A. Moslem, and T. Masoud. 2006. Comparison of Growth, Nutrition and Soil Properties of Pure and Mixed Stands of Populus Deltoids and *Alnus subcordata. Silva Fennica,* 40:27–35.

Scherer-Lorenzen, M., J. L. Bonilla, and C. Potvin. 2007. Tree Species Richness Affects Litter Production and Decomposition Rates in a Tropical Biodiversity Experiment. *Oikos,* 116:2108–2124.

Scherer-Lorenzen, M., C. Potvin, J. Koricheva, B. Schmid, A. Hector, Z. Bornik, G. Reynolds, and E.-D. Schulze. 2005. "The Design of Experimental Tree Plantations for Functional Biodiversity Research." In *Forest Diversity and Function: Temperate and Boreal System,* Ecological Studies 176, ed. M. Scherer-Lorenzen, C. Körner, and E.-D. Schulze, pp. 347–376. Berlin-Heidelberg: Springer-Verlag.

Schlapfer, F., and B. Schmid. 1999. Ecosystem Effects of Biodiversity: A Classification of Hypotheses and Exploration of Empirical Results. *Ecological Applications,* 9:893–912.

Schmid, B., A. Hector, M. A. Huston, P. Inchausti, I. Nijs, P. W. Leadley, and D. Tilman, 2002. "The Design and Analysis of Biodiversity Experiments." In *Biodiversity and Ecosystem Functioning: Synthesis and Perspectives,* ed. M. Loreau, S. Naeem, and P. Inchausti, pp. 61–78. New York: Oxford University Press.

Shiels, A. B. 2006. Leaf Litter Decomposition and Substrate Chemistry of Early Successional Species on Landslides in Puerto Rico. *Biotropica,* 38:348–353.

Siemann, E., D. Tilman, J. Haarstad, and M. E. Ritchie. 1998. Experimental Test of the Dependence of Arthropod Diversity on Plant Diversity. *American Naturalist,* 152:738–750.

Singh, J. S. 2002. The Biodiversity Crisis: A Multifaceted Review. *Current Science,* 82:638–647.

Srivastava, D. S., and M. Vellend. 2005. Biodiversity-Ecosystem Function Research: Is It Relevant to Conservation? *Annual Review in Ecology and Evolution Systems,* 36:267–294.

Stanley, W., and F. Montagnini. 1999. Biomass and Nutrient Accumulation in Pure and Mixed Plantation of Indigenous Tree Species Grown on Poor Soils in the Humid Tropics of Costa Rica. *Forest Ecology and Management,* 113:91–103.

Strong, D. R., J. H. Lawton, and T. R. E. Southwood. 1984. *Insects on Plants.* Cambridge, MA: Harvard University Press.

Systat Software I. 2002. Systat for Windows. Richmond, CA.

Tilman, D. 1999. The Ecological Consequences of Changes in Biodiversity: A Search for General Principles. *Ecology,* 80:1455–1474.

Vieira, S., P. B. Camargo, D. Selhorst, R. da Silva, L. Hutyra, J. Q. Chambers, I. F. Brown, N. Higuchi, J. dos Santos, S. C. Wofsy, S. E. Trumbore, and L. A. Martinelli. 2004. Forest Structure and Carbon Dynamics in Amazonian Tropical Rain Forests. *Oecologia,* 140:468–479.

Vojtech, N., and Y. Basset. 2005. Host Specificity of Insect Herbivores in Tropical Forests. *Proceedings of the Royal Society, B-Biological Sciences,* 272:1083–1090.

Wardle, D. A. 2002. "Plant Species Control of Soil Biota Processes." In *Communities and Ecosystems,* vol. 34, ed. D. A. Wardle, pp. 392. Princeton, NJ: Princeton University Press.

Wardle, D. A., K. I. Bonner, and K. S. Nicholson. 1996. Biodiversity and Plant Litter: Experimental Evidence Which Does Not Support the View That Enhanced Species Richness Improves Ecosystem Function. *Oikos,* 79:247–258.

Wardle, D. A., and P. Lavelle. 1997. "Linkages between Soil Biota, Plant Litter Quality and Decomposition." In *Driven by Nature: Plant Litter and Decomposition,* ed. G. Cadish and K. E. Giller, pp. 409. Wallingford, UK: CAB International.

Watt, A. 1997. Impact of Forest Management on Insect Abundance and Damage in a Lowland Tropical Forest in Southern Cameroon. *Journal of Applied Ecology,* 34:985–998.

Wright, J. 2002. Plant Diversity in Tropical Forests: A Review of Mechanisms of Species Coexistence. *Oecologia,* 130:1–14.

Zou, X., and G. Gonzalez. 1997. Changes in Earthworm Density and Community Structure during Secondary Succession in Abandoned Tropical Pastures. *Soil Biology and Biochemistry,* 29:627–629.

Biomass and Carbon Accumulation in Secondary Forests and Forest Plantations Used as Restoration Tools in the Caribbean Region of Costa Rica

William Fonseca,[1] Federico E. Alice,[1] Johan Montero,[1] Henry Toruño[1] and Humberto Leblanc[2]

ABSTRACT: Biomass and carbon accumulation were studied in secondary forests (5, 8 and 18 years old) and forest plantations of *Vochysia guatemalensis* and *Hyeronyma alchorneoides* at the EARTH University, located in the Caribbean zone of Costa Rica. Sampling plots, each of 500 m², were established in both forest ecosystems. The above- and belowground biomass, the necromass (litter and dead woody material) and the soil organic carbon were estimated in all plots. The carbon content in biomass was quantified by component. The highest carbon storage was found in plantations of *H. alchorneoides*, followed by plantations of *V. guatemalensis* and, lastly, secondary forests. The above- and belowground biomass and the necromass increased with age in the secondary forests and plantations. In contrast, the herbaceous biomass decreased with age in both ecosystems. The aboveground biomass stored between 11% and 17% of total carbon. Soil, at a depth of 30 cm, was the main carbon pool, accounting for 82.5–86.3% of the total ecosystem carbon.

Keywords: *aboveground biomass, belowground biomass, grasslands, native species, natural regeneration, necromass, soil organic carbon*

[1]*Instituto de Investigación y Servicios Forestales (INISEFOR), Universidad Nacional, Campus Omar Dengo 86-3000, Heredia, Costa Rica.*

[2]*Universidad EARTH, Guácimo, Limón, Costa Rica.*

Corresponding author: W. Fonseca (wfonseca@una.ac.cr).

INTRODUCTION

The rise in atmospheric temperature, which is caused by an increased concentration of greenhouse gases, is considered an indicator of global warming. Rising temperatures are currently being discussed throughout scientific, political, economic and environmental sectors. With the ratification of the Kyoto Protocol in 2005, a legal framework has been established for creating a global carbon market, providing opportunities for third world countries to obtain financial support to develop Land Use, Land Use Change and Forestry projects (LULUCF). Reforestation and afforestation, be it through active restoration (plantations) or passive restoration (natural regeneration), represents a valid option for Clean Development Mechanism (CDM) projects. In this context, it is necessary to develop reliable data for estimating the carbon capture capacity of forestry and agroforestry ecosystems in order to economically compensate stakeholders for the environmental services provided.

Twenty-five percent of Costa Rica is composed of forests listed under some type of conservation category (national parks and the like), plus 180,000 ha of primary commercial forests and 400,000 ha of secondary forests undergoing different successional stages. The country also has about 150,000 ha of forest plantations (reforestation rate of 3,500 ha yr^{-1}) and large areas of agroforestry (FAO, 2003). Costa Rica is among the most successful countries in its political and legislative environmental approaches, as well as with regard to the emerging carbon market and legislation on payments for environmental services. The country has available and suitable land for the development of carbon sequestration projects, where a vast quantity of land is underused or employed under land uses inferior to its capabilities.

Monitoring carbon storage in forest ecosystems is a priority in tropical countries that expect to participate in the carbon market (Sarlo et al., this volume). Therefore, the objective of this paper is the quantification of the amounts of carbon sequestered by plantations of *Vochysia guatemalensis* and *Hyeronyma alchorneoides*, as well as those quantities sequestered by secondary forests in the Caribbean region of Costa Rica.

MATERIALS AND METHODS

Study Area

This research was conducted at EARTH University, located in the Caribbean region of Costa Rica (N 10°10′; W 83°37′). The life zone is premontane very humid forest (Bolaños and Watson, 1993). Altitude varies between 64 and 95 m, with an annual precipitation of 3,464 mm distributed uniformly throughout the year and annual mean temperatures of 25.1°C. Study site soils are Typic Tropaquent associated with Tropic Fluvaquent: poorly developed, poorly drained and floodable at the lowlands (Gómez, 1986). This area has a regular topography, with slopes less than 5% and a ground water table above 90 cm deep.

Selection of Sampling Units

Forest plantations of *Vochysia guatemalensis, Hyeronyma alchorneoides* and secondary forests were selected with a wide age-range and located within similar soils, topography and climatic conditions. Sampling plots at secondary forests and forest plantations were established at EARTH University and "Las Delicias" farm on sites previously occupied by pasture lands (therefore including pasture lands as the control group; Tables 1 to 3).

Estimation of Biomass and Stored Carbon

Estimation of biomass and stored carbon were conducted following the methodology proposed by MacDicken (1997) with several modifications. A nested plot design was used to measure biomass components in different size plots.

Aboveground Biomass in Woody Components

In each forest plantation, woody components (DBH > 2.5 cm) were measured in a rectangular 500 m^2 plot. Diameter at breast height (DBH) was measured for every tree, while total height was measured for dominant trees only (the highest 10% of trees per plot). Biomass was estimated through the mean tree method, using the tree with average DBH to measure biomass. In the case of the 5-, 8- and 18-year-old secondary forests sampled, the DBH of all trees was measured. All trees in these 500 m^2 plots were identified for species. For the quantification of woody component biomass in secondary forests, the tree with mean DBH from every diameter class was used, considering as well the specie with the highest Importance Value Index (IVI), the IVI being the sum of abundance, frequency and dominance or basal area expressed in relative values (Krebs, 1985). Diameter classes were constructed using a 5 cm interval. All selected trees were harvested and separated in their components (stem, branches and leaves), each component was weighed in the field, and a 300 g sample was collected in order to determine its dry matter content (DMC) by drying the sample in an oven at 75°C for 72 h.

Biomass in Herbaceous Vegetation and Small Woody Material

Herbaceous vegetation and small woody material consisted of grasses, lianas, ferns, small plants and shrubs or tree seedlings with DBH < 2.5 cm. This component was quantified within a 1 × 1 m subplot in each corner of the 500 m^2 plot (grouping these four subplots into one sample for analysis), where all the material was harvested to ground level, weighed, and samples were collected for estimation of DMC.

Necromass

Necromass (all dead woody material found at ground surface) was divided into fine necromass (<2 cm) and large necromass (>2 cm) (Scott et al., 1992; Saldarriaga, 1994; Moran et al., 2000). Fine necromass was estimated from four 0.5 × 0.5 m subplots (grouping these four subplots into one sample for analysis), while large necromass was evaluated from one 5 × 5 m subplot, all distributed randomly throughout the 500 m^2 plot. In these subplots, the material was weighed and samples were collected to determine DMC.

Belowground Biomass

Methods proposed by Sierra et al. (2001) were used for the determination of belowground biomass. The root system was divided into large roots (diameters >5 mm) and fine roots (diameter < 5 mm). Large roots were determined through the excavation and extraction of the tree root system from the selected average trees. These roots were washed, later weighed and a sample was collected to determine DMC. For the quantification of fine roots, four sampling points were determined randomly within the plot (grouping these four sampling points into one sample for analysis), and a block of soil was then extracted with a 20 × 20 cm shovel at a 30 cm depth. These samples were taken to the laboratory where they were run through a 250 mm sieve with water so that soil, stones, and other organic residuals could be separated. The resulting fine roots were later dried and weighed.

Soil Organic Carbon

Stored soil carbon was quantified based on the soil's carbon content, bulk density and sampling depth (30 cm). Bulk density was determined through the cylinder method (MacDicken, 1997), taking four samples from the same random points used to determine belowground biomass.

Sample Preparation and Chemical Analysis

Samples from plant tissues (stem, branches, foliage, herbaceous vegetation, necromass and roots) were all dried in an oven for three days at a temperature of 60°C. They were ground to a particle size of 240 mm. Carbon content was determined following the methods by Pregl and Dumas (Bremner and Mulvaney, 1982) in an auto-analyzer (Perkin-Elmer series II, CHN/S 2400, Norway Co.).

RESULTS AND DISCUSSION

Secondary Forests

Total biomass, expressed in tons per hectare, in secondary forests ranged from 28.9 t ha[1] at 5 years of age to 65.9 t ha[1] at 18 years (Table 1); this represents an average fixation rate of 1.65 tC ha[-1] year-1. Hughes et al. (1999) found an average of 272.1 t ha[1] at 16 years of age. Corrales (1998) found 162.1 t ha[1] of biomass in 15-year-old secondary forests and 324.1 t ha[1] at primary forests in humid and very humid climates in Costa Rica. The soil stored an average of 86.3% of the total carbon in this system (Table 1), ranging from 72.9 tC ha[-1] at the baseline (pasture lands) to 159.6 tC ha[-1] at 18 years (Table 1). Cifuentes et al. (2004) found 93.5 tC ha[-1] of organic soil carbon, with similar values for primary and secondary forests. Feldpausch et al. (2004) found a soil carbon accumulation rate of 42 to 84 tC ha[-1] at a depth of 45 cm in secondary forests of 12 to 14 years of age. Valero (2004) indicates that carbon accumulation in biomass is faster than that of soils, but in soils carbon stability is greater.

Tree stems in secondary forests account for 7.4% of the total carbon of the ecosystem and around 38% of carbon stored in the biomass (9.6 t ha[-1]). Herbaceous vegetation, fine and

TABLE 1. Biomass (B) and stored carbon (C) per site and component (both expressed in t ha⁻¹) in secondary forests, Guácimo, Limón, Costa Rica (2006).

Site		EARTH-baseline*	Delicias-baseline*	EARTH-A38-P1	EARTH-A38-P2	Las Delicias P1	Las Delicias P2	EARTH-Las Ingas	EARTH-Los Brown
Age (years)				5	5	8	18	18	18
Mean DBH (cm)				4,9	5,0	9,7	9,0	10,7	10,2
Basal area (m² ha⁻¹)				7,9	7,4	9,2	20,8	18,1	16,7
Herbaceous vegetation	B	2,6	2,6	3,1	2,7	6,1	1,8	2,8	2,1
	C	1,2	1,1	1,3	1,1	2,6	0,7	1,1	0,9
Large necromass	B			0,0	1,2	3,8	7,4	0,5	0,0
	C			0,0	0,5	2,1	3,9	0,2	0,0
Fine necromass	B			2,1	6,2	3,3	4,4	6,8	8,7
	C			0,8	2,6	1,4	1,9	2,5	4,1
Stem	B			12,9	10,1	15,6	35,3	32,0	22,3
	C			5,5	4,6	7,2	14,9	15,2	10,0
Branches	B			3,4	2,9	4,0	10,5	10,5	9,2
	C			1,6	1,4	2,0	5,08	5,0	4,3
Leaves	B			1,4	0,7	2,7	2,5	1,6	4,1
	C			0,7	0,3	1,2	1,1	0,8	1,9
Large roots	B			5,1	4,3	8,6	7,3	8,1	8,2
	C			2,3	1,9	4,2	3,5	3,5	3,9
Fine roots	B			1,6	0,016	0,003	1,8	6,6	3,1
	C			0,6	0,006	0,0015	0,7	2,6	1,2
Total Biomass	B	2,6	2,6	29,6	28,1	44,1	71,1	68,9	63,8
	C	1,2	1,1	12,8	12,4	20,7	31,8	30,9	26,5
Soil organic carbon	C	73,0	73,0	72,3	104,8	73,5	101,6	159,6	114,6
Total Carbon	C	74,2	74,1	85,1	117,2	94,2	133,4	190,5	140,9

* = pasture lands

large necromass, branches, leaves and roots were the components that stored the least amount of carbon, varying from 0.85 tC ha^{-1} in fine roots to 3.2 tC ha^{-1} in branches and large roots (<2.8% of total carbon each, Table 1). Aboveground biomass plus large necromass stored 13.1% of the total system's carbon (18.5 t ha^{-1}). Brown and Lugo (1982) report between 2.6 and 3.8 tC ha^{-1} of necromass in primary forests, while Delaney et al. (1997) found between 2.4 and 5.2 tC ha^{-1}. Tanner (1980) reported necromass values of 3.8 to 6.0 tC ha^{-1} in Jamaican forests; while Raich (1983) found values of 0.7 tC ha^{-1} in Costa Rican secondary forests. Schroeder and Winjum (1995) found that litter represents only 5%–6% of total carbon in natural forests in Brazil, and Delaney et al. (1997) report values between 2.2 and 7.8% in Venezuela.

Biomass components and necromass, with the exception of herbaceous vegetation, show a tendency to increase with age (Table 1). Large necromass increased from 0.5 t ha^{1} at 5 years to 3.9 t ha^{-1} at 18 years; while litter ranged from 4.2 to 6.7 t ha^{1} and herbaceous vegetation decreased from 2.9 to 2.3 t ha^{1} during this same period. Fine necromass increased, possibly due to an increase in plant density and to pioneer plant mortality, such as Piperaceas. Vegetation growth causes canopy closure, reducing light radiation to the lower canopy level and therefore eliminating herbaceous vegetation. Herrera et al. (2001) found similar results in Colombia, with an exponential increase of necromass and a negative exponential decrease in herbaceous vegetation and small woody material with an increase in age. Belowground biomass (roots) accounted for 23.9% of total biomass (9.12 t ha^{-1}) and increase with age, ranging from 4.5 to 11.7 t ha^{-1} at 5 and 18 years, respectively (Table 1). Similar results have been found by other authors (Hertel et al., 2003, Jiménez and Arias, 2004).

Forest Plantations

Hyeronyma alchorneoides plantations had the greatest total carbon storage and fixation when compared to *V. guatemalensis*; they ranged from 75.4 to 201.8 tC ha^{-1} between 1 and 14 years of age. In *V. guatemalensis* plantations, carbon increased from 90.0 to 166.2 tC ha^{-1}, during the same time period (Tables 2 and 3). Similar to secondary forests, soil was the main carbon pool, with a contribution of an average of 86.3% of the total carbon content. Soil organic carbon increased with age, ranging from 66.5 and 73.0 tC ha^{-1} in recently established *V. guatemalensis* plantations and at control sites, respectively, to 143.1 tC ha^{-1} in 7-year-old plantations. In *H. alchorneoides*, this same component increased from 55.8 to 83.6 tC ha^{-1} in recently established sites and control, respectively, to 145.6 tC ha^{-1} in young plantations (5 years). Gutiérrez and Lopera (2001) found that soil (including roots) in *Pinus patula* plantations stored 53.3% of total carbon (139.5 tC ha^{-1}) up to a depth of 25 cm.

Stem biomass in *V. guatemalensis* plantations stored 10.7% (18.2 tC ha^{-1}) of total carbon content, while all aboveground biomass accounted for 17.0% from the total (21.4 tC ha^{-1}). Herbaceous vegetation biomass in *V. guatemalensis* plantations decreased with age, varying from 5.0 t ha^{-1} at initial ages to 0.5 t ha^{-1} at 7 years (Table 2). Similar to secondary forests, an increase in density and size in trees was responsible for the decrease of herbaceous biomass (Table 2). This trend was also observed in *H. alchorneoides* (Table 3). Understory components in *V. guatemalensis* functioned similar to those in secondary forests because their average carbon content represents >1.4% of the total carbon content. Similar proportions were found in leaves and branches (1.3% and 1.9% of total carbon, respectively). In *H. alchorneoides*, values

TABLE 2. Biomass (B) and stored carbon (C) per site and component (both expressed in t ha⁻¹) in forestry plantations of *Vochysia guatemalensis*, Guácimo, Limón, Costa Rica (2006).

Site		EARTH baseline *	Delicias baseline *	EARTH Tiro al blanco P1	EARTH Tiro al blanco P2	EARTH Puente-hamaca	Las Delicias P1	Las Delicias P2	EARTH el Cruce P1	EARTH el Cruce P2	EARTH la Bomba
Age (years)				1	1	1,5	4–5	4–5	7	7	14
Mean DBH (cm)				2,7	0,0	2,9	23,3	20,3	28,7	30	30,7
Basal area (m² ha⁻¹)				0,4	0,0	0,5	26,6	19,2	28,9	21,4	36,8
Herbaceous	B	2,6	2,6	2,3	7,7	5,0	0,9	2,0	0,4	0,5	9,3
vegetation	C	**1,2**	**1,2**	**0,9**	**3,2**	**2,0**	**0,4**	**0,9**	**0,2**	**0,2**	**3,8**
Large necromass	B			0,0	0,0	0,0	3,1	2,8	6,4	17,3	7,5
	C			**0,0**	**0,0**	**0,0**	**1,4**	**1,3**	**3,1**	**8,1**	**3,4**
Fine necromass	B			0,0	0,0	0,0	9,4	3,6	5,0	7,6	7,1
	C			**0,0**	**0,0**	**0,0**	**3,4**	**1,3**	**1,8**	**2,6**	**2,7**
Stem	B			0,6	0,09	1,0	45,8	35,0	82,8	41,0	106,2
	C			**0,3**	**0,04**	**0,5**	**20,9**	**19,2**	**35,7**	**17,1**	**51,7**
Branches	B			0,56	0,08	0,2	10,7	6,8	23,7	9,7	8,3
	C			**0,24**	**0,03**	**0,1**	**4,8**	**2,9**	**10,3**	**5,0**	**3,6**
Leaves	B			1,0	0,16	0,5	4,9	5,5	12,7	3,8	2,2
	C			**0,4**	**0,09**	**0,2**	**1,9**	**2,4**	**5,4**	**1,7**	**0,9**
Large roots	B			0,28	0,06	0,11	12,2	15,0	39,5	18,1	27,0
	C			**0,12**	**0,025**	**0,05**	**5,6**	**6,9**	**19,6**	**7,8**	**12,1**
Fine roots	B			1,8	0,56	1,3	2,1	0,7	3,9	2,9	2,6
	C			**0,7**	**0,21**	**0,5**	**0,8**	**0,26**	**1,4**	**1,4**	**1,0**
Total Biomass	B	2,6	2,6	6,6	8,7	8,2	89,3	71,5	174,4	101,0	170,2
	C	**1,2**	**1,2**	**2,7**	**3,6**	**3,4**	**39,2**	**35,2**	**77,5**	**43,9**	**79,2**
Soil organic carbon	C	73,0	73,0	107,2	66,5	95,3	100,2	132,4	117,9	143,1	87,0
Total Carbon	C	74,2	74,2	109,9	70,1	98,7	139,4	167,6	195,4	187,0	166,1

* = pasture lands

TABLE 3. Biomass (B) and stored carbon (C) per site and component (both expressed in t ha⁻¹) in forestry plantations of *Hyeronyma alchorneoides*. Guácimo, Limón, Costa Rica (2006).

Site		EARTH baseline*	Delicias baseline*	EARTH Vivero	EARTH Tiro al blanco P1	EARTH Tiro al blanco P2	EARTH Puente hamaca	Las Delicias	EARTH el Cruce	EARTH Papelera	EARTH Pozo Azul	EARTH Cruce Reserva	EARTH la Bomba
Age (years)				0,5	1	1	1,5	4–5	7	7	7	8	14
Mean DBH (cm)				0,0	2,3	2,6	2,	10,7	16,6	17,4	12,6	15,6	22,9
Basal area (m² ha⁻¹)				0,0	0,3	0,4	0,5	6,4	8,3	10,6	5,3	7,4	21,6
Herbaceous vegetation	B	2,6	2,6	1,7	1,5	2,8	3,7	0,6	2,8	2,2	7,0	3,9	0,2
	C	1,2	1,2	0,7	0,8	1,3	1,6	0,3	1,1	1,0	3,0	1,6	0,1
Large necromass	B			0,0	0,0	0,0	0,0	5,6	3,9	8,8	1,1	1,6	2,7
	C			0,0	0,0	0,0	0,0	2,6	1,8	1,5	0,5	0,7	1,3
Fine necromass	B			0,0	6,0	0,0	0,0	12,5	7,7	4,9	5,4	5,6	10,6
	C			0,0	2,5	0,0	0,0	5,3	3,3	2,2	2,8	2,3	5,0
Stem	B			0,07	0,4	0,6	0,3	16,3	34,5	36,9	14,3	19,1	116,3
	C			0,03	0,2	0,27	0,16	7,9	16,9	16,6	6,1	10,8	59,7
Branches	B			0,0	0,04	0,3	0,05	4,3	5,5	6,1	7,2	15,7	14,4
	C			0,0	0,02	0,13	0,02	2,0	2,6	3,5	2,9	6,8	6,6
Leaves	B			0,06	0,2	0,5	0,3	4,3	3,2	3,9	2,7	3,1	4,5
	C			0,03	0,08	0,26	0,14	1,8	1,5	2,0	1,5	2,0	2,3
Large roots	B			0,04	0,09	0,6	0,15	11,2	10,6	11,5	4,0	5,5	19,5
	C			0,02	0,04	0,3	0,07	4,6	6,9	6,3	2,0	2,6	9,5
Fine roots	B			0,2	2,4	0,2	1,1	4,0	2,9	2,5	3,5	2,4	7,5
	C			0,09	0,93	0,09	0,42	1,6	1,16	1,2	1,7	1,0	3,0
Total Biomass	B	2,6	2,6	2,0	10,6	5,0	5,5	58,8	71,2	76,7	45,2	57,8	175,8
	C	1,2	1,2	0,9	4,6	2,4	2,4	26,1	35,3	36,8	20,5	27,8	87,5
Soil organic carbon	C	84,0	83,2	73,0	89,5	55,8	90,0	145,6	96,2	118,5	112,6	92,4	114,5
Total Carbon	C	85,2	84,4	73,9	94,1	58,2	92,4	171,7	131,5	155,3	133,1	120,2	201,8

* = pasture lands

were quite similar, with less than 1.2% for herbaceous vegetation, 1.5% for large necromass and 1.8% for fine necromass. Branches accounted for 1.9%, leaves for 0.8% and stems for 7.2% of the total carbon content. Fine root biomass increased with age, but had a low contribution to total carbon content (0.6% and 0.8% of total carbon for *V. guatemalensis* and *H. alchorneoides*, respectively; Tables 2 and 3).

Increases in biomass and carbon have not yet been largely studied for forestry plantations of *V. guatemalensis* and *H. alchorneoides*. Stanley and Montagnini (1999) report biomass values of 35.9 and 27.3 t ha^{-1} in 3.5-year-old *H. alchorneoides* plantations and 3-year-old *V. guatemalensis* plantations, respectively. Shepherd and Montagnini (2001) report a biomass of 102.2 and 48.1 t ha^{-1} in a 5-year-old *V. guatemalensis* plantation and a 6-year-old *H. alchorneoides* plantation, respectively. Redondo-Brenes and Montagnini (2006) reported a biomass of 104.6 and 88.9 t ha^{-1} in a 13-year-old *V. guatemalensis* plantation and a 12-year-old *H. alchorneoides* plantation, respectively. Gutiérrez and Lopera (2001) found that litter and herbaceous vegetation in *Pinus patula* plantations store between 0.5% and 1.7% and between 0% and 3.2% of the total carbon, respectively. Gamarra (2001) found a 4% of total carbon in litter at eucalyptus plantations in Peru.

In *V. guatemalensis*, large and fine necromass increased with the plantation's age, with the highest values found immediately after silvicultural treatments (Table 2). Large necromass increased from 0 t ha^{-1} in recently established plantations to 7.5 t ha^{-1} at 14 years; while fine necromass increased from 0 to 7.1 t ha^{-1} during that same period. Different results were also found in plantations of the same age, such was the case for Delicias P1 and P2 and at EARTH el Cruce P1 and P2 (Table 2). These differences are due to contrasting tree densities, which strongly determine necromass deposition since *V. guatemalensis* is self-pruning. Necromass and litter also increased with plantation age in *H. alchorneoides* (Table 3). The effect of silvicultural treatments, such as thinning at the La Papelera site and pruning at Las Delicias, significantly increased large necromass (8.8 and 5.6 tC ha^{-1}; Table 3).

CONCLUSION

Biomass and stored carbon in secondary forests, *Vochysia guatemalensis* and *Hieronyma alchorneoides* plantations increased with age in all components, with the exception of herbaceous biomass. Secondary forests, on average, stored a total of 154.9 tC ha^{-1} at 18 years of age; while forest plantations of *V. guatemalensis* and *H. alchorneoides* reached 166.2 and 201.8 tC ha^{-1} at 14 years of age, respectively. The greatest increase in carbon storage was found in *H. alchorneoides* plantations, followed by *V. guatemalensis* and finally, secondary forests.

The soil stored between 82.5% and 86.3% of the total carbon content. Herbaceous vegetation and necromass stored the least amount of carbon. Stem biomass stored between 7.2% and 10.7% of the total carbon content, therefore constituting an important component in carbon sequestration projects.

ACKNOWLEDGMENTS

The authors thank Carlos Sandí for all his assistance and cooperation. We also thank Delio Zamora, Carlos Mata and Minor Cubillo for their collaboration during fieldwork and all of EARTH's University employees for their unconditional support.

The Spanish version for this paper can be found at:

Fonseca, W.; Alice, F.; Toruño, J.; Leblanc. 2008. Acumulación de biomasa y carbono en bosques secundarios y en plantaciones forestales de Vochysia guatemalensis e Hieronyma alchornoides en el Caribe de Costa Rica. Agroforestería en las Américas no. 46:57-63.

REFERENCES

Bolaños, R. A., and V. C. Watson. 1993. Mapa ecológico de Costa Rica según el sistema de clasificación de zonas de vida del mundo de Holdridge. San José, CR, Centro Científico Tropical.

Bremner, J. M., and C. Mulvaney. 1982. Carbon, Inorganic Nitrogen. In *Methods of Soil Analysis: Chemical and Microbiological Properties* 2 ed., ed. A. Page, R. Miller, and D. Keeney, pp. 552, 673–682. Madison, WI: American Society of Agronomy.

Brown, S., and A. Lugo. 1982. The Storage and Production of Organic Matter in Tropical Forest and Their Role in the Global Carbon Cycle. *Biotropica,* 14:164–187.

Cifuentes, M., J. Jobse, V. Watson, and B. Kauffman. (2004). Determinación del carbono total en suelos de diferentes tipos de uso suelo de la tierra a lo largo de una gradiente climática en Costa Rica. San Jose, Costa Rica. Available at http://www.una.ac.cr/inis/docs/suelos/VicWat.pdf. (accessed 25 July 2008).

Corrales, L. 1998. Estimación de la cantidad de carbono almacenado y captado (masa aérea) en el Corredor Biológico Mesoamericano de Costa Rica. PROARCA/CAPAS/CCAD/USAID.

Delaney, M., S. Brown, E. Lugo, A. Torres, and N. Bello. 1997. The Distribution for Organic Carbon in Major Components of Forest Located in Five Life Zones of Venezuela. *Journal of Tropical Ecology,* 13:697–708.

FAO. 2003. *Costa Rica frente al cambio climático. Serie centroamericana de bosques y cambio climático.* Comisión Centroamericana de Ambiente y Desarrollo, San José (Costa Rica). FAO: San José, Costa Rica.

Feldpausch, T. R., M. A. Rondon, E. C. Fernandes, S. J. Riha, and E. Wandelli. 2004. Carbon and Nutrient Accumulation in Secondary Forests Regenerating on Pastures in Central Amazonia. *Ecological Applications,* 14(4):164–176.

Gamarra, J. 2001. Estimación del contenido de carbono en plantaciones de Eucalyptus globulus Labill, en Junin, Perú. Valdivia, Chile. *Simposio Internacional Medición y Monitoreo de la captura de carbono en Ecosistemas Forestales*, 18–20 de Octubre de 2001.

Gómez, L. D. 1986. *Vegetación de Costa Rica: Vegetación y Clima de Costa Rica.* Volumen 1. UNED: San José, Costa Rica.

Gutiérrez, V. H., and J. Lopera. 2001. Metodología para la cuantificación de existencias y flujo de carbono en plantaciones forestales. Valdivia, Chile. *Simposio Internacional Medición y Monitoreo de la Captura de Carbono en Ecosistemas Forestales*, 18 al 20 de Octubre del 2001.

Herrera, M. A., J. del Valle, and S. Alonso. 2001. Biomasa de la vegetación arbórea y leñosa pequeña y necromasa en bosques tropicales primarios y secundarios de Colombia. Valdivia, Chile. *Simposio Internacional Medición y Monitoreo de la Captura de Carbono en Ecosistemas Forestales*, 18 al 20 de Octubre del 2001.

Hertel, D., C. Lueschner, and D. Hölscher. 2003. Size and Structure of Fine Root System in Old-Growth and Secondary Tropical Montane Forests (Costa Rica). *Biotropica,* 35(2):143–153.

Hughes, R. F., J. B. Kauffman, and V. J. Jaramillo. 1999. Biomass, Carbon and Nutrient Dynamics of Secondary Forest in a Humid Tropical Region of Mexico. *Ecology,* 80(6):1882–1907.

Jiménez, C., and D. Arias. 2004. *Distribución de biomasa y densidad de raíces finas en una gradiente sucesional de bosques en la zona norte de Costa Rica*. Cartago, Costa Rica: Instituto Tecnológico de Costa Rica.

Krebs, J. C. 1985. *Ecología: Estudio de distribución y abundancia*. Segunda edición. México: Ed. Harla.

MacDicken, K. 1997. A Guide to Monitoring Carbon Storage in Forestry and Agroforestry Projects. Forest Carbon Monitoring Program. Winrock International Institute for Agricultural Development (WRI). http://www.winrock.org/REEP/PUBSS.html.

Moran, J. A., M. G. Barker, A. J. Moran, P. Becker, and S. M. Ross. 2000. A Comparison of the Soil Water, Nutrient Status, and Litterfall Characteristics of Tropical Heath and Mixed Dipterocarp Forest Sites in Brunei. *Biotropica,* 32:2–13.

Raich, J. 1983. Effects of Forest Conversion of the Carbon Budget of a Tropical Soil. *Biotropica*, 15(3):177–184.

Redondo-Brenes, A., and F. Montagnini. 2006. Growth, Productivity, Aboveground Biomass, and Carbon Sequestration of Pure and Mixed Native Tree Plantations in the Caribbean Lowlands of Costa Rica. *Forest Ecology and Management,* 232:168–178.

Saldarriaga, J. G. 1994. Recuperación de la selva de tierra firme en el alto río Negro Amazonía Colombiana-Venezolana. Estudios de la Amazonia Colombiana V. Bogotá, Tropenbos, Colombia, 1994. Tomo V, Pág. 201.

Scott, D. A., J. Proctor, and J. Thomoson. 1992. Ecological Studies on a Lowland Evergreen Rain Forest on Maracá Island, Romaira, Brazil. II. Litter and Nutrient Cycling. *Journal of Ecology,* 80:705–717.

Schroeder, P., and J. Winjum. 1995. Assessing Brazil's Carbon Budget: I. Biotic Carbon Pools. *Forest Ecology and Management,* 75:77–86.

Shepherd, D., and F. Montagnini. 2001. Aboveground Carbon Sequestration Potential in Mixed and Pure Tree Plantations in the Humid Tropics. *Journal of Tropical Forest Science,* 13(3):450–459.

Sierra, C., J. del Valle, and S. Orrego. 2001. Ecuaciones de biomasa de raíces y sus tasas de acumulación en bosques sucesionales y maduros tropicales en Colombia. Valdivia, Chile. *Simposio Internacional Medición y Monitoreo de la Captura de Carbono en Ecosistemas Forestales*, 18 al 20 de Octubre del 2001.

Stanley, W., and F. Montagnini. 1999. Biomass and Nutrient Accumulation in Pure and Mixed Plantations of Indigenous Tree Species Grown on Poor Soils in the Humid Tropics of Costa Rica. *Forest Ecology and Management,* 113:91–103.

Tanner, E. 1980. Studies on the Biomass and Productivity in a Series of Montane Rain Forest of Jamaica. *Journal of Ecology,* 68:573–588.

Valero, E. 2004. *El ciclo del carbono en el sector forestal: "Los bosques como sumidereos de carbono: una necesidad para cumplir en el protocolo de Kyoto."* Madrid, España: Universidad del Vigo.

Climate Change and Snow Depth Impacts on Vegetation at the Great Basin Desert— Sierra Nevada Ecotone

Michael E. Loik,[1] Holly Alpert[1] and Alden B. Griffith[1,2]

ABSTRACT: General circulation model scenarios of future precipitation patterns are highly uncertain in terms of future patterns of snow depth and snowmelt timing. Because snowfall provides the majority of annual soil water recharge in many western high-elevation North American ecosystems, this study tested hypotheses about the linkages of snow depth to soil water content, recruitment, and plant diversity at the ecotone between the Great Basin Desert shrub-steppe and Sierra Nevada conifer forest. Snow depth was manipulated using eight long-term (>50 yr) snow fences near Mammoth Lakes, Mono County, California, USA. Snow depth, soil moisture content, and vegetation characteristics were assessed in response to increased and decreased snow depth. Snow depth on increased-depth ("+snow") plots was about twice that of ambient-depth plots, and about 2.2 times that for decreased-depth ("−snow") plots. Soil water content on +snow plots was double that on ambient and −snow plots. Species richness and diversity were lowest on −snow plots. There was a shift from co-dominance of the Great Basin Desert shrubs *Artemisia tridentata* and *Purshia tridentata* on ambient and +snow plots, to dominance of *P. tridentata* on −snow plots. Frequency of seedlings and dead plants of these two species were consistent with patterns of shifting dominance. The frequency of saplings of the Sierra Nevada conifers *Pinus contorta* and *Pi. jeffreyi* were related to snow depth,

[1] *Department of Environmental Studies, 1156 High St., University of California, Santa Cruz, CA 95064, USA.*
[2] *Wellesley College, Department of Biological Sciences, 106 Central Street, Wellesley, MA 02481, USA.*
Corresponding author: M. Loik (mloik@ucsc.edu).

but more so to microsites associated with the two shrub species. Results suggest that an increased frequency of ENSO events will not impact the diversity of the Great Basin Desert—Sierra Nevada ecotone as much as would a decrease of snow depth and earlier melt timing, a possibility envisioned by an increasing number of climate model scenarios.

Keywords: *Artemisia tridentata, ENSO, Pinus contorta, Pinus jeffreyi, Purshia tridentata, soil water, snowfall*

INTRODUCTION

Precipitation patterns have changed for many regions of the United States since the early 1900s (Vorosmarty and Sahagian, 2000; Groisman et al., 2004), a trend that many general circulation models (GCMs) envision will continue as a result of anthropogenic activities (Mearns et al., 1995; Hughen et al., 1998; Easterling et al., 2000; NAST, 2000; IPCC, 2001; Hayhoe et al., 2004). Wintertime snowfall and melt timing particularly impact organisms and ecosystems, especially in montane regions (Frei and Robinson, 1999; Loik et al., 2004a; Saito et al., 2004; Frei and Gong, 2005; Lapp et al., 2005). Snow intermittently covers about 2×10^6 km^2 of the western United States, and dramatic shifts in snow climate (depth, extent, and melt timing) may be in store for about 55% of terrestrial North America in the future. Changes in winter precipitation patterns will have critical impacts on distributions of western ecosystems because snow provides 70% to 90% of their annual hydrologic input, and snowmelt is a key timing event for the onset of growth for many species. Snow-covered regions of the western United States are typically composed of grasslands, arid and semiarid cold deserts, shrublands, and forests from the Cascades and Sierra Nevada, across the Intermountain semideserts/deserts, the Nevada/Utah ranges, the Colorado Plateau, and the Rocky Mountains. For the western United States, this encompasses ca. 1.73×10^8 ha, or about 22% of the coterminous United States (Bailey et al., 1994). Based on a modest net primary productivity for these range and forest lands of 100 gC m^{-2} yr^{-1}, this area corresponds to a net production of approximately $3.46 \times 1,011$ gC yr^{-1}. Thus, snowfall is a major abiotic resource input for western U.S. vegetation communities.

Numerous studies have shown that snow depth and snowmelt timing are fundamental determinants of community structure and ecosystem processes particularly for western, northern, and high-elevation North America (Kudo, 1991; Galen and Stanton, 1993, 1995; Kudo, 1993; Harte et al., 1995; Kudo et al., 1999; Stinson, 2004). Snowfall and snowmelt trigger the onset and magnitude of many ecological processes, as well as the quantity and quality of stream flow available for agricultural, domestic, and ecosystem needs (Regonda et al., 2005; Zierl and Bugmann, 2005). In concert with changing precipitation patterns, warmer atmospheric temperatures will truncate the amount of time that snow covers terrestrial ecosystems, largely through earlier onset and increased rate of melt (Harte et al., 1995; Stinson, 2004; Wahren et al., 2005).

Microscale variation in snow depth and snow bank persistence helps to determine the spatial and temporal patterns of recruitment and mortality that characterize plant populations

(Galen and Stanton, 1993, 1995; Walker et al., 1993, 1995; Harte and Shaw, 1995; Price and Waser, 1998; Loik et al., 2004b; Yamagishi et al., 2005). The timing of snowmelt marks the onset of seasonal plant growth and subsequent levels of abiotic stress that may impact reproductive effort. For example, earlier onset of melting driven by experimental infrared warming in the Rocky Mountains, Colorado, USA, causes certain species to be vulnerable to a longer drought in the subsequent summer (Karsh and MacIver, this volume; Shaw et al., 2000; Saavedra et al., 2003). Certain species may be vulnerable to physiological damage at snowmelt due to the loss of niveal insulation (Loik et al., 2004b), and altered levels of flower, fruit and seed production later in time (Lambrecht et al., 2007).

Although there is an increasing understanding of the role of precipitation pulses in driving plant community diversity and ecosystem processes (for example, Breshears et al., 1997a, 1997b, 1998; Breshears and Barnes, 1999; Schwinning and Sala, 2004; Schwinning et al., 2004) the ecohydrology of snow (that is, snow depth, snowpack persistence, melt timing) is far more complex than for rain (Loik et al., 2004a). Pulses of snowfall do not directly translate into soil moisture the same way as rainfall pulses, largely because of the time lag between snowfall and snowmelt infiltration into the soil, and because of the loss of snow pack water due to sublimation (Marks and Dozier, 1992; Groisman and Easterling, 1994; Cline, 1997; Hood et al., 1999; Cayan et al., 2001; Essery et al., 2003; French and Binley, 2004; Lee and Mahrt, 2004; Mahrt and Vickers, 2005; Murray and Buttle, 2005).

Few experimental global change studies incorporate different global climate change model/regional climate change model scenarios, have long-term histories, nor encompass large landscape spatial scales, especially for wintertime snow depth in high-elevation or high-latitude ecosystems. Many scenarios envision reduced Sierra Nevada snowfall and/or earlier melt timing (Snyder et al., 2002; Hayhoe et al., 2004; Barnett et al., 2008). On the other hand, there is little understanding of how future ENSO events (which deliver considerably large amounts of snowfall to the Sierra Nevada) may be affected by anthropogenic activities. Thus, there is a critical need to conduct experiments that quantify the effects of more than one direction or magnitude of precipitation change across large spatial and temporal scales. To meet this demand, this study compares in situ, long-term (>50 year), large-scale (that is, arrayed across 50 km of the landscape), manipulative simulations of increased vs. decreased snowfall on shrub and tree population processes. These two treatments serve as analogues of increased ENSO activity (increased snow, for example, HadCM2) and reduced snowfall and earlier melt timing (decreased snow, for example, HadCM3). This approach allows for an incorporation of climate model uncertainty into an ecological prediction framework, increasing the utility of these predictions for climate change adaptation (for example, fire risk), and conservation management (for example, invasive species control, ecological restoration) in this study region.

The objective of this study was to assess the response of soil water content, species richness, cover, recruitment and mortality to experimental manipulations of snow depth. This research is part of a broader study of the short- and long-term impacts of snow depth forcing on biodiversity, nutrient cycling, and the potential for carbon sequestration under increased and decreased winter precipitation regimes. Experiments were conducted in eastern California at the ecotone of the Great Basin Desert shrub land with the Sierra Nevada conifer forest ecosystems; such ecotones may be particularly sensitive to future changes in precipitation and other global changes (Bachelet et al., 2003; Neilson et al., 2005). The following hypotheses were tested: (1) soil moisture would be higher on research plots with greater snow depth—and lower on plots with reduced snow depth—compared to control, ambient-depth plots; (2) such changes in soil water availability would result in differences in species richness, cover, mortality, and seedling frequency on ambient-, increased-, and decreased-depth plots; and (3) recruitment microsites for two widespread Sierra Nevada conifer species would differ across snow depth treatments as well as recruitment site availability.

MATERIALS AND METHODS

Study Site

Experiments were conducted near Mammoth Lakes, Mono County, California, USA (37°38′54″N, 118°58′19″W, 2,400 m a.s.l.). The region is well known for its tectonic and volcanic activity (Farrar et al., 1985). The soils are derived from glacial and alluvial deposits, and volcanic material; soils are characterized as belonging to the Cozetica, Vitrandic Xerorthent, Cryopsammet, Haypress, and Torriothentic Haploxeroll families. There are no restrictive layers in the soil above 100 to 150 cm depth at each of the snow fence sites (Seney and Gallegos, 1995). The soils at all study sites have high rates of infiltration (15 to 50 cm h^{-1}), and gas and water permeability (Seney and Gallegos, 1995), and therefore there is very little to no surface runoff.

Experimental Design

The variation in response variables were compared (i.e., soil water content, species richness, cover, and diversity) as a function of three snow-depth treatments: (1) increased snow depth ("+snow"); (2) decreased snow depth ("–snow"); and (3) unmanipulated ("ambient") snow depth. Snow depth was increased and decreased by disruption of prevailing wintertime laminar wind flow using snow fences (Tabler, 1974). Snow fences create distinct spatial footprints of increased and decreased snow depth on the lee side of the fence, independent of the amount of ambient snowfall in a particular year. The location of increased and decreased snow depth is consistent from year to year.

Research plots were established in February, 2003 on either side of eight permanent snow fences situated adjacent to U.S. Highway 395 in the Inyo National Forest, Mono County, California. The snow fences were installed by the California Department of Transportation ("Caltrans") in the 1950s as part of road snow control efforts on U.S. 395. The snow fences occur over a 50 km transect along the west side of Hwy 395 from southeast of Mammoth Lakes, California, to east of June Lake, California. The fences are oriented approximately north-northwest to east-southeast, generally parallel to the direction of U.S. 395 at each site and perpendicular to the prevailing wind direction in winter. The fences range from 100 to 200 m away from the ditch of the highway. Seven of the snow fences are of the "Wyoming" type, 4 m in height, and 50% in porosity (Tabler, 1974); the other fence is made of metal slats and is 2.5 m tall with 50% porosity. The fences vary from 100 to more than 200 m in length; the 100 m long research plots (see below) were selected so that they were located on relatively flat terrain (slope <1%), and at locations where adjacent fences do not overlap one another. There has been no mowing or brush control along the fences, and the fences are maintained annually by Caltrans. Maintenance is conducted via access roads located at the immediate base of the fence on the immediate upwind side, where least damage to research plots could happen.

Based on snow depth patterns measured in January 2003 (and subsequently confirmed in 2004–2008), plots were established for each snow fence at locations corresponding to ambient snow depth, maximal increased depth, and maximal decreased depth ("ambient," "+snow," "–snow," respectively). Each plot was 5 m wide * 100 m long; the long axis of each plot is parallel to the snow fence. The ambient depth plots were centered 50 m upwind of the snow fences, the +snow plots were centered at the maximum measured snow depth downwind of the fences, and the –snow plots were located downwind of the fences where snow depth is ca. 80% of the upwind ambient depth. The altered snow depths on +snow and –snow plots serve as analogues of the HadCM2 and HadCM3 scenarios of increased (i.e., enhanced ENSO) and decreased (i.e., reduced snowfall and earlier melt timing) snow depth, respectively.

Meteorological Data

Historic snow depth from 1928 to 2005 was obtained for Mammoth Lakes, California, from the California Data Exchange Center of the California Department of Water Resources (http://cdec.water.ca.gov/). Monthly mean maximum and minimum air temperatures and precipitation data for 1994 to 2005 were obtained from the National Climatic Data Center (NCDC, http://www.ncdc.noaa.gov/oa/ncdc.html). Daily precipitation (i.e., rain and melted snow water equivalents), snowfall, and snow depth for Mammoth Lakes, California, were obtained from NCDC for the period from June 2004 to July 2005.

Soil Moisture

Soil moisture for 2004 was measured gravimetrically in April for one sample for each treatment at each fence. In September 2004, one 20 cm long Decagon ECH_2O dielectric aquameter probe (Decagon Devices, Pullman, WA, USA) was buried horizontally at 50 cm depth in each transect for all fences. ECH_2O probes were connected to RM5 data loggers set to record the voltage through the soil moisture probes at 12-hour intervals from September 2004 through September 2005.

Vegetation Surveys

Vegetation was characterized using a modification of methods used by the U.S. National Park Service for assessing fire risk (USDI National Park Service, 2003). Vegetation surveys were conducted using a 100 m long line transect along the middle of each plot for the control, increased snow depth, and decreased snow depth plots for all eight snow fences. For each itemized intercept along the length of the transect, a 0.5 cm wide survey pole was vertically placed next to the plant and transect tape, and with one end of the pole on the ground. The resulting intercept location was recorded as all plant species contacted by the pole, bare ground, and/or coarse litter (generally the dead stems and branches of *A. tridentata* or *P. tridentata*).

Diversity across treatments was computed as species richness (S) or using the Shannon Index, H´ = –p$_i$ > log p$_i$ (Begon et al., 1996). Cover was calculated as cover of species i = (sum of the intercept lengths for species i) / (sum of intercept lengths for all species). To compare the relative amounts of *A. tridentata* and *P. tridentata* cover independent of bare ground, relative percent cover was calculated as (sum of the intercept for *A. tridentata* or *P. tridentata*) / (sum of intercept lengths for all vegetation per transect). Mortality of *A. tridentata* and *P. tridentata* was computed as the cover of dead stems based on the total length of coarse litter intercepts.

Seedling recruitment for *A. tridentata* and *P. tridentata* was based on detailed searches of 2 × 100 m belt transects centered on the 100 m line transects described above. Detailed searches consisted of close examination of open, intershrub sites, as well as subcanopy sites under all individuals of *A. tridentata* and *P. tridentata* within the 200 m^2 belt for each treatment and snow fence. Recruitment for *Pinus contorta* and *Pi. jeffreyi* was conducted in a similar manner as for the two shrub species, except that the belt transect was 4 m wide × 100 m long to capture the greater spatial heterogeneity in their occurrence. Where possible (based on normality of data and independence of measurements), vegetation data were analyzed using a one factor (snow depth) analysis of variance.

RESULTS

Meteorology

There was considerable variation in historic patterns of snow depth for the period 1928 to 2005 (Figure 1A), with a mean of 1,368 mm and one standard deviation of 649 mm. Nine years had April snow depths one standard deviation below the mean, in contrast to 16 years with snow depth greater than one standard deviation above the mean. For the period 1993 to 2005, monthly average air temperatures ranged from a daytime maximum of 5°C and nighttime minimum of –8°C in January, to a daytime maximum of 27°C and nighttime minimum of 4°C in July (Figure 1B). For the same period, 84% of the average annual precipitation (= 636 mm) fell between October and April. Monthly snowfall is greatest from December to March, but is quite variable in all winter months (Figure 1C). In typical snowfall years, snow pack accumulation begins in October or November (Figure 1D), but can start in late September in El Niño years. Melting events can occur throughout the Winter and Spring due to episodic warm spells. The snow pack persists until March or April, but can be later in heavy snowfall years.

There were 49 precipitation events (rain + snow) for the typical snow year of 2003–2004, totaling 416 mm, whereas there were 70 events totaling 787 mm in the El Niño snow year of 2004–2005 (Figure 2A). Excluding rain, there were 34 snowfall events in 2003–2004, with the largest delivering 609 mm (Figure 2B). For 2004–2005, there were 51 snowfall events, with the largest event totaling 711 mm. Snow began to fall slightly earlier in 2004 (19 October) than in 2003 (1 November). There was an ephemeral snow-free period in December for both years. The duration of the snow-covered season was 146 days for 2003–2004, and 205 days for 2004–2005. The rate of snowmelt at the end of the season was 5.65 mm d^{-1} in 2003–2004 and 3.54 mm d^{-1} in 2004–2005. Snow had completely melted by 26 March in 2004, and by 12 May in 2005 (Figure 2C).

FIGURE 1. Climatology and meteorology of the study site in Mono County, California, USA. (A) Historic snow depth in April for 1928–2005. The solid horizontal line is the mean, the dashed horizontal lines are ±1 SD, and the dotted horizontal lines are ±2 SD. Data are from the California Data Exchange Center. (B) Mean monthly maximum (°C) and minimum (°C) air temperatures, and precipitation (bars) for 1993–2004 from the Mammoth USFS Ranger Station (via the NCDC). (C) Average monthly snow depth. Data are means ±1 SD, from the same source as in B. (D) The average number of days per month with at least 25.4 mm of snow on the ground. Data are means ±1 SD, from the same source as in B.

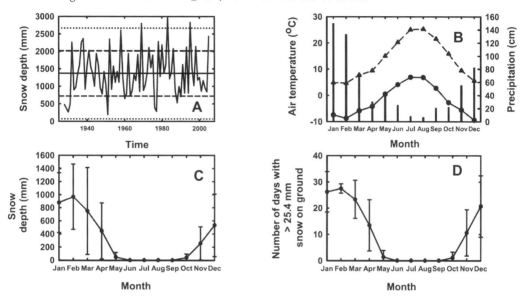

Snow Depth Manipulation and Soil Moisture

The intended effect of the snow fences was prominent (Figure 3A), and the location of maximal and minimal snow depth zones relative to the snow fences was consistent for the winters of 2003 to 2005. In 2004, snow depth averaged 70 cm (range 50 to 95 cm) on ambient plots, and was approximately twice as deep on +snow plots (Figure 3A). In contrast, snow depth at 50 m downwind of the snow fence was generally about 50% of the upwind depth (Figure 3A).

There were differences in snow depth on ambient, +snow and −snow plots when all fences were analyzed (Figure 3B). Snow depth was lower in 2004 compared to 2005 (Figure 3B). Snow depth for 2004 was greatest on +snow plots, but not different for −snow in comparison to ambient-depth plots. In contrast, snow depth for 2005 was highest on +snow plots, and lowest on −snow plots.

Soil Water Content

For July 2004, soil water content was 18 ± 3% for +snow plots, 14 ± 2% for ambient-depth plots, and 12 ± 0.1% for −snow plots. Soil water content for the winter and spring of 2004–2005 measured at 50 cm depth exhibited treatment and time-dependent differences (Figure 4). Soil water content increased in all treatments during a snowmelt event in November 2004 after the first snowfall of October. For −snow plots there were several ephemeral melt events

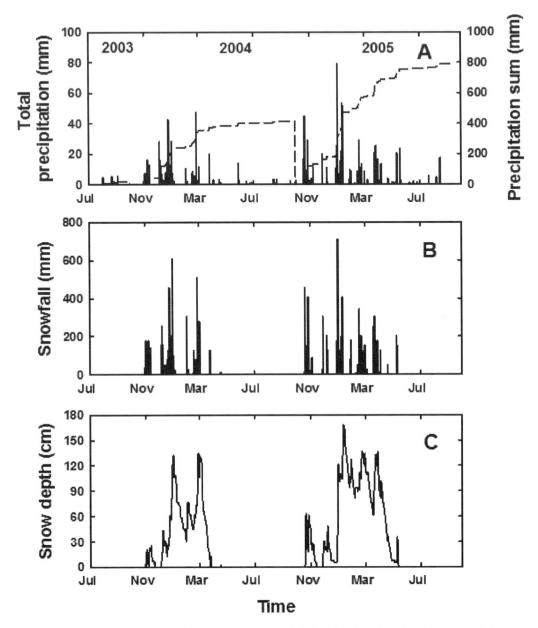

FIGURE 2. Precipitation for the hydrologic years 2003–2004 and 2004–2005. (A) Total precipitation per event (vertical bars) and precipitation sum for each year (dashed lines). (B) Snowfall per precipitation event. (C) Snow depth. Data were obtained from the Mammoth USFS Ranger Station (via the NCDC).

indicated by abrupt but short-lived increases in soil moisture, beginning in February and continuing through May. Patterns and magnitude of soil water content for ambient-depth plots were similar to those for −snow plots, but the rapid increases in soil moisture were not as pronounced. Soil moisture in −snow plots peaked and begin to decline several weeks before ambient-depth plots reached maximum soil moisture. The first melt event and soil moisture increase for +snow plots occurred in the first half of March. Overall, the maximum soil moisture on +snow plots after snowmelt was about twice that on ambient and −snow plots. Soil

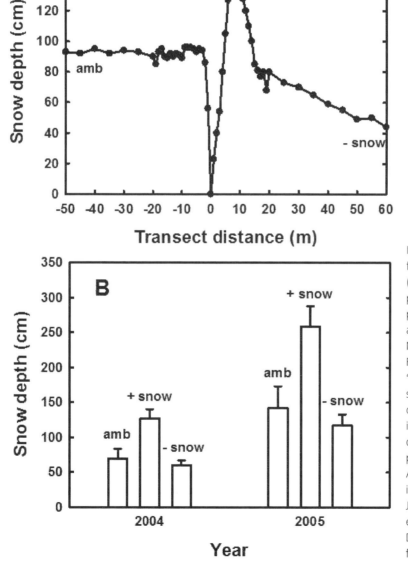

FIGURE 3. Snow fence forcing of snow depth. (A) Typical snow depth profile measured perpendicular to the long axis of "Deadman North" snow fence, February 2004. "amb," "+snow," and "−snow" show the location of ambient-depth, increased-depth and decreased-depth plots, respectively. (B) Average snow depth in January 2004 and January 2005 for ambient, +snow, and −plots. Data are means ±1 SD, for n = 8 snow fences.

moisture at this depth decreased more rapidly for +snow plots in late May in comparison to ambient and –snow plots. Overall, soil moisture at 50 cm depth was similar in magnitude and drying kinetics to the 0 to 20 cm depth (Figure 4, insets).

Vegetation

The shrubs *Artemisia tridentata* and *Purshia tridentata* were by far the most common species at all eight snow fence sites, but about 30% of the habitat is occupied by bare ground. The conifer trees *Pinus contorta* and *Pi. jeffreyi* are scattered in an apparent non-random fashion. Other vegetation recorded at the snow fences includes *Achnatherum thurberianum* (Poaceae) (Roemer & Schulties) Barkworth, *Carex rossii* (Cyperaceae), *Chrysothamnus naseosus* (Asteraceae) (Pallas) Britton, *Elymus elemoides* (Poaceae) (Raf.) Swezey, *Eriogonum spergulinum*, *E. umbellatum*, (Polygonaceae) Gray, *Gayophytum diffusum* (Onagraceae) (Torrey & A. Gray), *Leptodactylon pungens* (Polemoniaceae) (Torrey) Rydb. and *Lupinus lepidus* (Fabaceae) Douglas.

Diversity computed as the Shannon Index was significantly lower on –snow in comparison to ambient-depth and +snow plots (Figure 5A; one-way ANOVA F = 5.895, P = 0.0043). Species richness was lower on –snow compared to +snow but not ambient depth plots (Figure 5B; F = 3.124, P = 0.043).

Based on the total vegetation cover, *A. tridentata* and *P. tridentata* were roughly codominant on ambient-depth and +snow plots (Figure 6A). This is consistent with previous measurements at nearby sites. However, on –snow plots, relative percent cover for *A. tridentata* was about 35% while that for *P. tridentata* was approximately 65%. Dead stems of the two species can be distinctly identified revealing that there was greater cover of dead stems of *A. tridentata* on –snow compared to ambient-depth plots (Figure 6B). There was a nonsignificant trend toward higher cover of dead stems of *P. tridentata* on –snow compared to +snow and ambient-depth plots. The number of seedlings less than 10 cm tall for both shrub species was quite similar on both ambient and +snow plots, but was considerably lower for *A. tridentata* on –snow plots (Figure 6C).

Pinus contorta seedlings were more commonly observed on ambient-depth plots in comparison to +snow and –snow plots (Figure 7A). By contrast, *Pi. jeffreyi* seedlings were more common on –snow than ambient and +snow plots. When compared across microhabitats, *Pi. contorta* seedlings were equally frequent in open, intershrub sites and under the canopy of *P. tridentata*, and none were found under *A. tridentata* canopies (Figure 7B). For *Pi. jeffreyi*, quite a few more seedlings were found under the canopy of the two shrub species in comparison to open, intershrub microsites.

DISCUSSION

Results indicate that greater snow depth led to higher soil moisture content on +snow (i.e., increased-depth) plots following snowmelt, in comparison to ambient and –snow plots. The effect of the snow fences on snow depth in +snow and –snow plots is consistent across years, resulting in an increase (on +snow plots) and decrease (on –snow plots) in snow depth and soil moisture year in and year out. Presumably the observed differences in plant diversity have resulted at least in part to these long-term effects of snow depth forcing and soil moisture

FIGURE 4. Mean volumetric soil moisture content at 50 cm depth between September 2004 and September 2005. Data are means for n = 8 snow fences per transect. Graphs are plotted with the same scale on the ordinate to facilitate comparison. Insets: Mean volumetric soil moisture content at 50 cm (solid line) and 0 to 20 cm depth (dashed line).

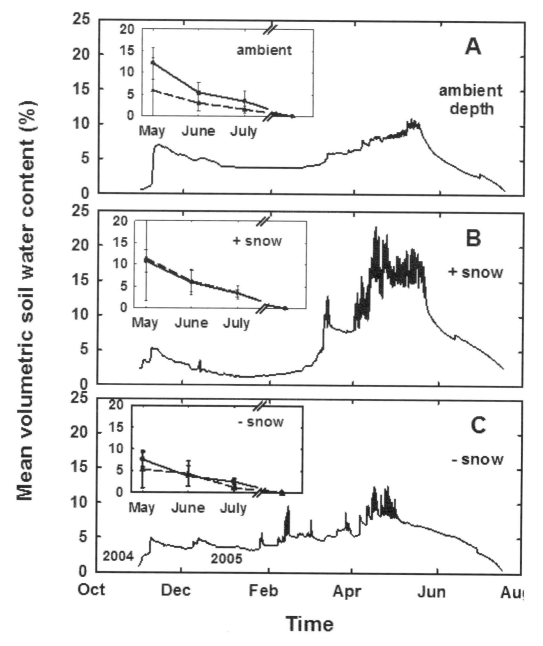

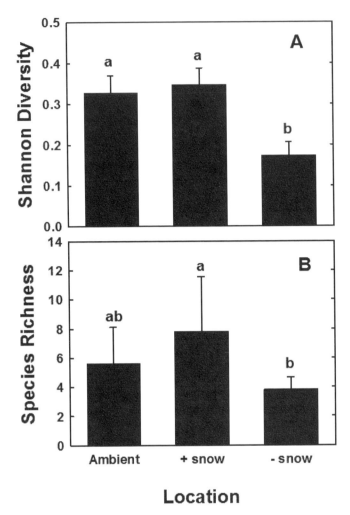

FIGURE 5. Vegetation community characteristics. (A) Shannon diversity in 2003 for locations at each snow fence with ambient, +snow, and −snow depth. (B) Species richness in 2003 for locations at each snow fence with ambient, +snow, and −snow depth. For both graphs, data are means ±1 SD, for n = 8 snow fences.

availability on plant recruitment and mortality. Diversity is lower, and dominance by the two Great Basin Desert shrub species has shifted on plots where snow has melted earlier and soil moisture has been lower for the past several decades. The patterns of *A. tridentata* recruitment and mortality across long-term snow depth treatments are consistent with the shift in dominance. Higher mortality and reduced recruitment of *A. tridentata* on −snow plots may underlie the dominance shift toward *P. tridentata*, for which mortality and recruitment seem to be unaffected by the reduced snow depth. The spatial distribution of the Sierra Nevada conifer species *Pinus contorta* and *Pi. jeffreyi* appears to also be affected by snow depth. Yet, their distribution also seems to be related to the presence of *A. tridentata* and *P. tridentata* canopies, suggesting an interaction between snow depth and recruitment sites. Despite these observations, there is considerable interannual variation in snowfall, snow depth, and snowmelt timing at this site such that water stress in some low snowfall years may be subsidized by higher levels of snowfall and snowmelt infiltration in other years. Thus, the great temporal variability in soil moisture content across time complicates our ability to link seedling recruit-

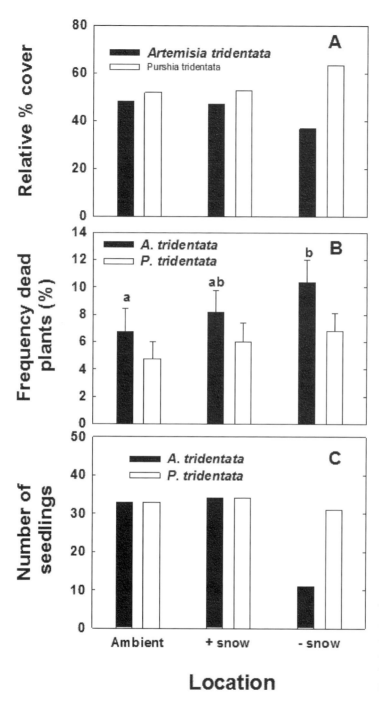

FIGURE 6. (A) Relative percent cover (cover as a total amount of vegetation) for *Artemisia tridentata* and *Purshia tridentata* as a function of snow depth treatment. Data are composites for all eight fences. (B) Cover of dead stems of *A. tridentata* and *P. tridentata* as a function of snow depth treatment. Data are means ±1 SD, for n = 8 snow fences. For *A. tridentata*, bars with different letters have P < 0.05; differences are not significant for *P. tridentata*. (C) Number of seedlings as a function of snow-depth treatment. Data are composites for all eight fences.

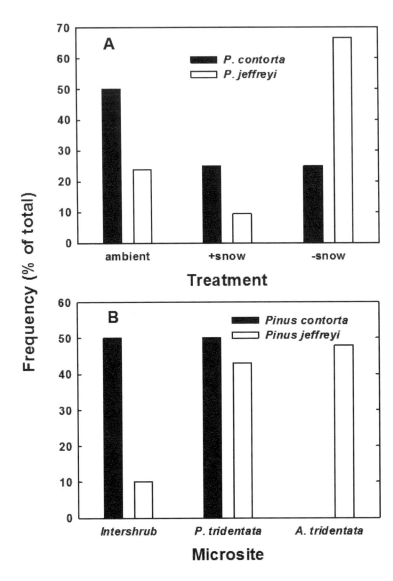

FIGURE 7. (A) Frequency of seedlings of *Pinus contorta* and *Pinus jeffreyi* as a function of snow depth. Data are composites for all eight fences. (B) Frequency of *Pinus contorta* and *Pinus jeffreyi* seedlings as a function of microhabitat type. "Intershrub" refers to open space between the canopies of *A. tridentata* and *P. tridentata*. Data are composites for all eight fences.

ment and adult plant mortality at the Great Basin Desert–Sierra Nevada ecotone to specific years of adequate or inadequate soil water availability.

Most research on plant responses to global change has focused on physiological and phenological changes by adult individuals, despite the fact that seedling recruitment, which is acutely affected by microclimate, is a primary factor influencing plant community composition (Smith and Nowak, 1990; Davis and Zabinski, 1992; Smith et al., 1997; Neilson et al., 2005). For plants in the western United States, seedling establishment is often the life history stage that is most vulnerable to stress because of drought near the soil surface. Seed germination and seedling establishment patterns determine community dynamics over both large and small spatial and temporal scales. In particular, species-specific recruitment patterns will be key in determining the rate and extent of changes in species' distributions in response to global change, especially at sensitive ecotones (Neilson, 1993; Woodward et al., 1998; Bachelet et al., 2003). For many species, recruitment is determined by the ability to survive the rigors of the microenvironment adjacent to the soil surface (Nobel, 1999; Smith and Nowak, 1990; Larcher, 1995), and for the arid and semiarid regions of the western United States drought is an important recruitment filter (Loik et al., 2004a). In general, an increase in precipitation alone would be expected to enhance both seed germination and seedling growth rates (Weltzin and McPherson, 1997; Germaine and McPherson, 1998), whereas the opposite would be expected for decreased precipitation. However, plant community responses are likely to be complex and difficult to predict. Establishment is often facilitated by the presence of adult nurse plants, whose canopies modify the abiotic environment and ameliorate the stressful conditions for seedlings (Niering et al., 1963; de Jong and Klinkhamer, 1988; Franco and Nobel, 1989; Callaway et al., 1996; Holmgren et al., 1997). In some cases, the benefit of growing in the canopy of another plant is offset by the cost of competition for certain resources, such as water or soil nutrients (Franco and Nobel, 1989; Callaway et al., 1996). Despite much research on the mechanistic factors associated with facilitation of seedling establishment, still little is known about how climate change will influence the association between seedlings and their nurse plants (Callaway and Walker, 1997; Holmgren et al., 1997).

The top 20 cm of the soil surface dried earlier on −snow plots compared to ambient and +snow plots, and in comparison to the 50 cm deep soil layer. Assuming that surface soil moisture content is correlated with snowmelt timing (i.e., earlier-melting plots dry earlier), this could result in a phenological shift across treatments similar to that reported for infrared (IR) warming effects in the Colorado Rocky Mountains (Harte et al., 1995). In the present study there were differences in the timing of vegetative growth as well as flower and fruit production for both Great Basin Desert shrub species. In particular, growth and flowering were about 10 to 14 days advanced for both shrub species on −snow plots in comparison to those on +snow and ambient-depth plots. Snow persisted in patches on ambient and +snow plots in early May whereas the −snow plots were completely melted. Observations on growth timing are consistent with results from Perfors et al. (2003), who found that growth rate for *A. tridentata* is higher for plants exposed to overhead IR warming, due to the effects of heaters on snowmelt date rather than on soil temperature or moisture. Perhaps earlier snowmelt at the snow fence sites allows shrubs a slightly longer growing season. By contrast, a study that combined IR warming with snow removal treatments showed that timing of flowering for *A. tridentata* was not readily explained by snowmelt timing, soil moisture, or temperature, although ten other conspecific species were affected (Dunne et al., 2003). Clearly, not all species respond in a similar manner to differences in snow depth, snowmelt timing, soil moisture, or temperature (Kudo, 1991, 1993; Price and Waser, 1998; Kudo et al., 1999; Pop et al., 2000; Schutz, 2002; Van Wijk et al., 2003).

Our understanding of ecological responses to altered precipitation—particularly for snowfall, snow pack evolution, and snowmelt timing—is constrained by GCM/RCM resolution and uncertainty (MacCracken et al., 2003; Miller et al., 2003). The magnitude, timing, and the sign of the envisioned changes vary across model scenarios. For example, HadCM2 envisions up to 100% more precipitation during winter by 2090–2099 for snow-covered regions of the Sierra Nevada of California (VEMAP Members, 1995). In contrast, HadCM3 envisions a 15% to 20% decrease. RCMs incorporate greater spatial and temporal resolution, but there is still considerable uncertainty. For instance, RegCM2.5 suggests reduced snow accumulation in all major hydrologic regions of California (Snyder et al., 2002, 2004), portending major implications for diversity and functioning of terrestrial ecosystems (Diffenbaugh and Sloan, 2004). Comparisons of the Parallel Climate Model with HadCM3 suggest anywhere from no change to dramatic reductions in seasonal snowfall for the Sierra Nevada, yet April snow pack (a critical metric for ecosystem water availability and agricultural and other human uses) may be reduced by 30% to 90% (Hayhoe et al., 2004).

In conclusion, soil water content during late spring and summer did reflect to a large degree snow depth in the previous winter, as evidenced by snow fence forcing of depth and melt timing. The present results suggest that snow depth impacts on soil moisture will alter the co-dominance of the Great Basin Desert shrubs *Artemisia tridentata* and *Purshia tridentata* at snow-dominated locations where they overlap with the Sierra Nevada conifer forests of eastern California. Moreover, such shifts may alter the availability of recruitment sites for the Sierra Nevada conifer species *Pinus contorta* and *Pi. jeffreyi*, highlighting complex interactions between the spatial and temporal patterns of physical and biotic factors at this site. This study capitalized on long-term, concurrent increases and decreases in snow depth that capture potential scenarios of enhanced ENSO behavior or reduced snowfall and earlier snowmelt.

It is still far from clear whether and how the frequency and intensity of ENSO events will change under continued anthropogenic forcing of atmospheric radiative gas concentrations (Yeakley et al., 1994; Frei and Robinson, 1999; Peterson et al., 2000; Bachelet et al., 2003; Bell et al., 2004; Fekete et al., 2004; Groisman et al., 2004; Lee et al., 2004; Saito et al., 2004; Ashrit et al., 2005; Frei and Gong, 2005). Moreover, an increasing number of ensemble modeling assessments envision earlier snowmelt (Barnett et al., 2008), pointing toward shifting shrub dominance and conifer recruitment site availability.

Nevertheless, despite uncertainties in future GCM, RCM and ENSO prognostications, this research concerning the responses to climate change demonstrate the need to consider impacts of contrasting snowfall scenarios on ecohydrologic characteristics, population processes, and thereby long-term impacts on diversity at ecotones.

ACKNOWLEDGMENTS

The Student Challenge Awards Program of Earthwatch, the M. Theo Kearney Foundation for Soil Science and the National Institute for Climate Change Research (U.S. Department of Energy) provided financial support. The Valentine Eastern Sierra UC Natural Reserve provided housing and other logistics. The enthusiastic assistance of Amy Concilio, Lucy Lynn, Ian Gillespie, Hally Andersen and numerous Earthwatch SCAP volunteers helped make this research possible.

REFERENCES

Ashrit, R. G., A. Kitoh, and S. Yukimoto. 2005. Transient Response of ENSO-Monsoon Teleconnection in MRI-CGCM2.2 Climate Change Simulations. *Journal of the Meteorological Society of Japan,* 83:273–291.

Bachelet, D., R. P. Neilson, T. Hickler, R. J. Drapek, J. M. Lenihan, M. T. Sykes, B. Smith, S. Sitch, and K. Thonicke. 2003. Simulating Past and Future Dynamics of Natural Ecosystems in the United States. *Global Biogeochemical Cycles* 17(2):14-1–14-21.

Bailey, R. G., P. E. Avers, T. King, and W. H. McNab. 1994. *Ecoregions and Subregions of the United States.* Washington, DC: USDA Forest Service.

Barnett, T. P., D. W. Pierce, H. G. Hidalgo, C. Bonfils, B. D. Santer, T. Das, G. Bala, A. W. Wood, T. Nozawa, A. A. Mirin, D. R. Cayan, and M. D. Dettinger. 2008. Human-Induced Changes in the Hydrology of the Western United States. *Science,* 319:1080–1083.

Begon, M., J. L. Harper, and C. R. Townsend. 1996. *Ecology: Individuals, Populations, and Communities,* 3rd ed. Cambridge, MA: Blackwell Science Ltd.

Bell, J. L., L. C. Sloan, and M. A. Snyder. 2004. Regional Changes in Extreme Climatic Events: A Future Climate Scenario. *Journal of Climate,* 17:81–87.

Breshears, D. D., and F. J. Barnes. 1999. Interrelationships between Plant Functional Types and Soil Moisture Heterogeneity for Semiarid Landscapes within the Grassland/Forest Continuum: A Unified Conceptual Model. *Landscape Ecology,* 14:465–478.

Breshears, D. D., J. W. Nyhan, C. E. Heil, and B. P. Wilcox. 1998. Effects of Woody Plants on Microclimate in a Semiarid Woodland: Soil Temperature and Evaporation in Canopy and Intercanopy Patches. *International Journal of Plant Sciences,* 159:1010–1017.

Breshears, D. D., O. B. Myers, S. R. Johnson, C. W. Meyer, and S. N. Martens. 1997a. Differential Use of Spatially Heterogeneous Soil Moisture by Two Semiarid Woody Species: *Pinus edulis* and *Juniperus monosperma. Journal of Ecology,* 85:289–299.

Breshears, D. D., P. M. Rich, F. J. Barnes, and K. Campbell. 1997b. Overstory-Imposed Heterogeneity in Solar Radiation and Soil Moisture in a Semiarid Woodland. *Ecological Applications,* 7: 1201–1215.

Callaway, R. M., E. H. DeLucia, D. Moore, R. Nowak, and W. D. Schlesinger. 1996. Competition and Facilitation: Contrasting Effects of *Artemisia tridentata* on *Pinus ponderosa* versus *P. monophylla. Ecology,* 77:2130–2141.

Callaway R. M., and L. R. Walker. 1997. Competition and Facilitation: A Synthetic Approach to Interactions in Plant Communities. *Ecology,* 78:1958–1965.

Cayan, D. R., S. A. Kammerdiener, M. D. Dettinger, J. M. Caprio, and D. H. Peterson. 2001. Changes in the Onset of Spring in the Western United States. *Bulletin of the American Meteorological Society,* 82:399–415.

Cline, D. W. 1997. Snow Surface Energy Exchanges and Snowmelt at a Continental, Mid-Latitude Alpine Site. *Water Resources Research,* 33:689–701.

Davis M. B., and C. Zabinski 1992. "Changes in Geographical Range Resulting from Greenhouse Warming Effects on Biodiversity in Forests." In *Global Warming and Biological Diversity*, ed. R. L. Peters and T. L. Lovejoy, pp. 298–308. New Haven, CT: Yale University Press.

de Jong, T. J., and P. G. L. Klinkhamer. 1988. Population Ecology of the Biennials *Cirsium vulgare* and *Cynoglossum officinale* in a Coastal Sand-Dune Area. *Journal of Ecology,* 76:366–382.

Diffenbaugh, N. S., and L. C. Sloan. 2004. Mid-Holocene Orbital Forcing of Regional-Scale Climate: A Case Study of Western North America Using a High-Resolution RCM. *Journal of Climate,* 17:2927–2937.

Dunne, J. A., J. Harte, and K. J. Taylor. 2003. Subalpine Meadow Flowering Phenology Responses to Climate Change: Integrating Experimental and Gradient Methods. *Ecological Monographs,* 73:69–86.

Easterling, D. R., G. A. Meehl, C. Parmesan, S. A. Changnon, T. R. Karl, and L. O. Means. 2000. Climate Extremes: Observations, Modeling, and Impacts. *Science* 289:2068–2074.

Essery, R., J. Pomeroy, J. Parviainen, and P. Storck. 2003. Sublimation of Snow from Coniferous Forests in a Climate Model. *Journal of Climate,* 16:1855–1864.

Farrar, C. D., M. L. Sorey, S. A. Rojstaczer, C. J. Janik, R. H. Mariner, and T. L. Winnett. 1985. Hydrologic and Geochemical Monitoring in Long Valley Caldera, Mono County, California, 1982–1984. *U.S. Geological Survey Water-Resources Investigations Report* 85-4183. Sacramento, CA.

Fekete, B. M., C. J. Vorosmarty, J. O. Roads, and C. J. Willmott. 2004. Uncertainties in Precipitation and Their Impacts on Runoff Estimates. *Journal of Climate,* 17:294–304.

Franco, A. C., and P. S. Nobel. 1989. Effect of Nurse Plants on the Microhabitat Growth of Cacti. *Journal of Ecology,* 77:870–886.

Frei, A., and G. Gong. 2005. Decadal to Century Scale Trends in North American Snow Extent in Coupled Atmosphere-Ocean General Circulation Models. *Geophysical Research Letters,* 32.

Frei, A., and D. A. Robinson. 1999. Northern Hemisphere Snow Extent: Regional Variability 1972–1994. *International Journal of Climatology,* 19:1535–1560.

French, H., and A. Binley. 2004. Snowmelt Infiltration: Monitoring Temporal and Spatial Variability Using Time-Lapse Electrical Resistivity. *Journal of Hydrology,* 297:174–186.

Galen, C., and M. L. Stanton. 1993. Short-Term Responses of Alpine Buttercups to Experimental Manipulations of Growing-Season Length. *Ecology,* 74:1052–1058.

———. 1995. Responses of Snowbed Plant-Species to Changes in Growing-Season Length. *Ecology,* 76:1546–1557.

Germaine, H. L., and G. R. McPherson. 1998. Effects of Timing of Precipitation and Acorn Harvest Date on Emergence of *Quercus emoryi. Journal of Vegetation Science,* 9:157–160.

Groisman, P. Y., and D. R. Easterling. 1994. Variability and Trends of Total Precipitation and Snowfall over the United States and Canada. *Journal of Climate,* 7:184–205.

Groisman, P. Y., R. W. Knight, T. R. Karl, D. R. Easterling, B. M. Sun, and J. H. Lawrimore. 2004. Contemporary Changes of the Hydrological Cycle over the Contiguous United States: Trends Derived from In Situ Observations. *Journal of Hydrometeorology,* 5:64–85.

Harte, J., and R. Shaw. 1995. Shifting Dominance within a Montane Vegetation Community: Results of a Climate-Warming Experiment. *Science,* 267:876–880.

Harte, J., M. S. Torn, F. R. Chang, B. Feifarek, A. P. Kinzig, R. Shaw, and K. Shen. 1995. Global Warming and Soil Microclimate: Results from a Meadow-Warming Experiment. *Ecological Applications,* 5:132–150.

Hayhoe, K., D. Cayan, C. B. Field, P. C. Frumhoff, E. P. Maurer, N L. Miller, S. C. Moser, S. H. Schneider, K. N. Cahill, E. E. Cleland, L. Dale, R. Drapek, R. M. Hanemann, L. S. Kalkstein, J. Lenihan, C. K. Lunch, R. P. Neilson, S. C. Sheridan, and J. H. Verville. 2004. Emissions Pathways, Climate Change, and Impacts on California. *Proceedings of the National Academy of Sciences of the United States of America,* 101:12422–12427.

Holmgren, M., M. Scheffer, and M. A. Huston. 1997. The Interplay of Facilitation and Competition in Plant Communities. *Ecology,* 78:1966–1975.

Hood, E., M. Williams, and D. Cline. 1999. Sublimation from a Seasonal Snowpack at a Continental, Mid-Latitude Alpine Site. *Hydrological Processes,* 13:1781–1797.

Hughen, K. A., J. T. Overpeck, S. J. Lehman, M. Kashgarian, J. Southon, L. C. Peterson, R. Alley, and D. M. Sigman. 1998. Deglacial Changes in Ocean Circulation from an Extended Radiocarbon Calibration. *Nature,* 391:65–68.

[IPCC] Intergovernmental Panel on Climate Change. 2001. *Climate Change 2001: The Scientific Basis.* Contribution of Working Group I to the Third Assessment Report of the IPCC. Cambridge, UK: Cambridge University Press.

Kudo, G. 1991. Effects of Snow-Free Period on the Phenology of Alpine Plants Inhabiting Snow Patches. *Arctic and Alpine Research,* 23:436–443.

———. 1993. Relationship between Flowering Time and Fruit-Set of the Entomophilous Alpine Shrub *Rhododendron aureum* (Ericaceae), Inhabiting Snow Patches. *American Journal of Botany,* 80:1300–1304.

Kudo, G., U. Nordenhall, and U. Molau. 1999. Effects of snowmelt timing on leaf traits, Leaf Production, and Shoot Growth of Alpine Plants: Comparisons along a Snowmelt Gradient in Northern Sweden. *Ecoscience,* 6:439–450.

Lambrecht, S. C., M. E. Loik, D. W. Inouye, and J. Harte. 2007. Reproductive and Physiological Responses to Simulated Climate Warming for Four Subalpine Species. *New Phytologist,* 173:121–134.

Lapp, S., J. Byrne, I. Townshend, and S. Kienzle. 2005. Climate Warming Impacts on Snowpack Accumulation in an Alpine Watershed. *International Journal of Climatology,* 25:521–536.

Larcher, W. 1995. *Physiological Plant Ecology.* Berlin: Springer Verlag.

Lee, S., A. Klein, and T. Over. 2004. Effects of the El Nino-Southern Oscillation on Temperature, Precipitation, Snow Water Equivalent and Resulting Streamflow in the Upper Rio Grande River Basin. *Hydrological Processes,* 18:1053–1071.

Lee, Y. H., and L. Mahrt. 2004. An Evaluation of Snowmelt and Sublimation over Short Vegetation in Land Surface Modelling. *Hydrological Processes,* 18:3543–3557.

Loik, M. E., D. D. Breshears, W. K. Lauenroth, J. Belnap. 2004a. A Multi-Scale Perspective of Water Pulses in Dryland Ecosystems: Climatology and Ecohydrology of the Western USA. *Oecologia,* 141:269–281.

Loik, M. E., C. J. Still, T. E. Huxman, and J. Harte. 2004b. In Situ Photosynthetic Freezing Tolerance for Plants Exposed to a Global Warming Manipulation in the Rocky Mountains, Colorado, USA. *New Phytologist,* 162:331–341.

MacCracken, M. C., E. J. Barron, D. R. Easterling, B. S. Felzer, and T. R. Karl. 2003. Climate Change Scenarios for the U.S. National Assessment. *Bulletin of the American Meteorological Society,* 84:1711.

Mahrt, L., and D. Vickers. 2005. Moisture Fluxes over Snow with and without Protruding Vegetation. *Quarterly Journal of the Royal Meteorological Society,* 131:1251–1270.

Marks, D., and J. Dozier. 1992. Climate and Energy Exchange at the Snow Surface in the Alpine Region of the Sierra Nevada. 2. Snow Cover Energy-Balance. *Water Resources Research,* 28:3043–3054.

Mearns, L. O., F. Giorgi, L. McDaniel, and C. Shields. 1995. Analysis of Daily Variability of Precipitation in a Nested Regional Climate Model: Comparison with Observations and Doubled CO_2 Results. *Global and Planetary Change,* 10:55–78.

Miller, A. J., M. A. Alexander, G. J. Boer, F. Chai, K. Denman, D. J. Erickson, R. Frouin, A. J. Gabric, E. A. Laws, M. R. Lewis, Z. Y. Liu, R. Murtugudde, S. Nakamoto, D. J. Neilson, J. R. Norris, J. C. Ohlmann, R. I. Perry, N. Schneider, K. M. Shell, and A. Timmermann. 2003. Potential Feedbacks between Pacific Ocean Ecosystems and Interdecadal Climate Variations. *Bulletin of the American Meteorological Society,* 84:617–633.

Murray C. D., and J. M. Buttle. 2005. Infiltration and Soil Water Mixing on Forested and Harvested Slopes during Spring Snowmelt, Turkey Lakes Watershed, Central Ontario. *Journal of Hydrology,* 306:1–20.

[NAST] National Assessment Synthesis Team. 2000. *Climate Change Impacts on the United States: The Potential Consequences of Climate Variability and Change.* U.S. Global Change Research Program. New York: Cambridge University Press.

Neilson, R. P. 1993. Transient Ecotone Response to Climatic-Change: Some Conceptual and Modeling Approaches. *Ecological Applications,* 3:385–395.

Neilson R. P., L. F. Pitelka, A. M. Solomon, R. Nathan, G. F. Midgley, J. M. V. Fragoso, H. Lischke, and K. Thompson. 2005. Forecasting Regional to Global Plant Migration in Response to Climate Change. *Bioscience,* 55:749–759.

Niering, W. A., R. H. Whittaker, and C. H. Lowe. 1963. The Saguaro: A Population in Relation to Environment. *Science,* 142:15–23.

Nobel, P. S. 1999. *Physicochemical and Environmental Plant Physiology.* San Diego, CA: Academic Press.

Perfors, T., J. Harte, and S. E. Alter. 2003. Enhanced Growth of Sagebrush (*Artemisia tridentata*) in Response to Manipulated Ecosystem Warming. *Global Change Biology,* 9:736–742.

Peterson, D. H., R. E. Smith, M. D. Dettinger, D. R. Cayan, and L. Riddle. 2000. An Organized Signal in Snowmelt Runoff over the Western United States. *Journal of the American Water Resources Association,* 36:421–432.

Pop, E. W., S. F. Oberbauer, and G. Starr. 2000. Predicting Vegetative Bud Break in Two Arctic Deciduous Shrub Species, *Salix pulchra* and *Betula nana. Oecologia,* 124:176–184.

Price, M. V., and N. M. Waser. 1998. Effects of Experimental Warming on Plant Reproductive Phenology in a Subalpine Meadow. *Ecology,* 79:1261–1271.

Regonda, S. K., B. Rajagopalan, M. Clark, and J. Pitlick. 2005. Seasonal Cycle Shifts in Hydroclimatology over the Western United States. *Journal of Climate,* 18:372–384.

Saavedra, F., D. W. Inouye, M. V. Price, and J. Harte. 2003. Changes in Flowering and Abundance of *Delphinium nuttallianum* (Ranunculaceae) in Response to a Subalpine Climate Warming Experiment. *Global Change Biology,* 9:885–894.

Saito, K., T. Yasunari, and J. Cohen. 2004. Changes in the Sub-Decadal Covariability between Northern Hemisphere Snow Cover and the General Circulation of the Atmosphere. *International Journal of Climatology,* 24:33–44.

Schutz, W. 2002. Dormancy Characteristics and Germination Timing in Two Alpine Carex Species. *Basic and Applied Ecology,* 3:125–134.

Schwinning, S., and O. E. Sala. 2004. Hierarchy of Responses to Resource Pulses in Arid and Semiarid Ecosystems. *Oecologia,* 141:211–220.

Schwinning, S., O. E. Sala, M. E. Loik, and J. R. Ehleringer. 2004. Thresholds, Memory, and Seasonality: Understanding Pulse Dynamics in Arid/Semi-Arid Ecosystems. *Oecologia,* 141:191–193.

Seney, J. P., and J. A. Gallegos. 1995. *Soil Survey of Inyo National Forest, West Area, California.* Vallejo, CA: United States Forest Service, Pacific Southwest Region.

Shaw, M. R., M. E. Loik, and J. Harte. 2000. Gas Exchange and Water Relations of Two Rocky Mountain Shrub Species Exposed to a Climate Change Manipulation. *Plant Ecology,* 146:197–206.

Smith, S. D., R. K. Monson, and J. E. Anderson. 1997. *Physiological Ecology of North American Desert Plants.* New York: Springer-Verlag.

Smith, S. D., and R. S. Nowak. 1990. "Ecophysiology of Plants in the Intermountain Lowlands." In *Plant Biology of the Basin and Range*, ed. C. B. Osmond, L. F. Pitelka, and G. F. Hidy, pp. 179–241. New York: Springer-Verlag.

Snyder, M. A., J. L. Bell, L. C. Sloan, P. B. Duffy, and B. Govindasamy. 2002. Climate Responses to a Doubling of Atmospheric Carbon Dioxide for a Climatically Vulnerable Region. *Geophysical Research Letters* 29(11).

Snyder, M. A., L. C. Sloan, and J. L. Bell. 2004. Modeled Regional Climate Change in the Hydrologic Regions of California: A CO_2 Sensitivity Study. *Journal of the American Water Resources Association,* 40:591–601.

Stinson, K. A. 2004. Natural Selection Favors Rapid Reproductive Phenology in *Potentilla pulcherrima* (Rosaceae) at Opposite Ends of a Subalpine Snowmelt Gradient. *American Journal of Botany,* 91:531–539.

Tabler, R. D. 1974. Design of a Watershed Snow Fence System and First-Year Snow Accumulation. 39th Western Snow Conference. Colorado State University, Fort Collins, Colorado.

USDI National Park Service. 2003. *Fire Monitoring Handbook.* Boise, ID: Fire Management Program Center, National Interagency Fire Center.

Van Wijk, M. T., M. Williams, J. A. Laundre, and G. R. Shaver. 2003. Interannual Variability of Plant Phenology in Tussock Tundra: Modelling Interactions of Plant Productivity, Plant Phenology, Snowmelt and Soil Thaw. *Global Change Biology,* 9: 743–758.

VEMAP Members. 1995. Vegetation/Ecosystem Modeling and Analysis Project: Comparing Biogeography and Biogeochemistry Models in a Continental-Scale Study of Terrestrial Ecosystem Responses to Climate Change and CO_2 Doubling. *Global Biogeochemical Cycles,* 9:407–437.

Vorosmarty, C. J., and D. Sahagian. 2000. Anthropogenic Disturbance of the Terrestrial Water Cycle. *Bioscience,* 50:753–765.

Wahren, C. H. A., M. D. Walker, and M. S. Bret-Harte. 2005. Vegetation Responses in Alaskan Arctic Tundra after Eight Years of a Summer Warming and Winter Snow Manipulation Experiment. *Global Change Biology,* 11:537–552.

Walker, D. A., J. C. Halfpenny, M. D. Walker, and C. A. Wessman. 1993. Long-Term Studies of Snow Vegetation Interactions. *Bioscience,* 43:287–301.

Walker, M. D., R. C. Ingersoll, and P. J. Webber. 1995. Effects of Interannual Climate Variation on Phenology and Growth of Two Alpine Forbs. *Ecology,* 76:1067–1083.

Weltzin, J. F., and G. R. McPherson. 1997. Spatial and Temporal Soil Moisture Resource Partitioning by Trees and Grasses in a Temperate Savanna, Arizona, USA. *Oecologia,* 112:156–164.

Woodward, F. I., M. R. Lomas, and R. A. 1998. Vegetation Climate Feedbacks in a Greenhouse World. *Philosophical Transactions of the Royal Society of London B Biological Sciences* 353:2939.

Yamagishi, H., T. D. Allison, and M. Ohara. 2005. Effect of Snowmelt Timing on the Genetic Structure of an *Erythronium grandiflorum* Population in an Alpine Environment. *Ecological Research,* 20:199–204.

Yeakley, J. A., R. A. Moen, D. D. Breshears, and M. K. Nungesser. 1994. Response of North American Ecosystem Models to Multiannual Periodicities in Temperature and Precipitation. *Landscape Ecology,* 9:249–260.

Zierl B., and H. Bugmann. 2005. Global Change Impacts on Hydrological Processes in Alpine Catchments. *Water Resources Research,* 41:1–13.

Appendix: Panama Statement

From the Climate Change and Biodiversity Symposium in the Americas

PANAMA CITY, PANAMA
FEBRUARY 2008

BIODIVERSITY

Biodiversity means the variability among all living organisms, including diversity within species, between species and of ecosystems. This diversity is the result of four billion years of evolution.

Biodiversity supports human societies ecologically, economically, culturally and spiritually. Despite their importance, ecosystems are being degraded, and species and genetic diversity are declining at an alarming rate. This is due to the impact of a number of forcing agents, including a changing climate, growing human populations and increasing resource consumption.

The decline in biodiversity is now recognized as one of the most serious environmental issues facing humanity. A global goal has been defined: to reduce the rate of loss of biodiversity by 2010 (www.biodiv.be/convention/2010targetllinks/lnk-world/int_conv/cbd/2010_target).

ADVICE AND GUIDANCE

1. Support the establishment of a climate change and biodiversity monitoring and information network throughout the Americas. This network will provide:
    a) a transect of scientific expertise and/or on-the-ground monitoring across chemical, climate and ecological gradients to allow for unique investigations into the impacts of climate change on forest biodiversity and improve our understanding of its adaptive capacity;
    b) mechanisms for the sharing and communicating of information, data and science on climate change and biodiversity;

   c) training on monitoring tools and methodologies, analysis techniques, data management and verification tools needed to support adaptation options and decision making; and

   d) a network secretariat, located at the Smithsonian Institution, for overall coordination of the monitoring network.

2. Support the establishment of a climate change and biodiversity research network throughout the America. This network will provide:

   a) integrated research on climate change and biodiversity, including new global and regional climate change models, hazards and extremes, and other human pressures impacting forest biodiversity;

   b) expert scientific advice on adaptation options and opportunities to reduce the impacts of climate change on biodiversity;

   c) an exchange of scientists, environmental managers and community leaders to increase the scientific capacity, training and development of new study methods and monitoring techniques for both climate change and biodiversity; and

   d) interlinking of research groups, such as Environment Canada's Adaptation and Impacts Research Division and the Smithsonian Institution, by taking advantage of new scientific developments (e.g., www.cccsn.ca, www.hazards.ca). To support the next generation of model development, transfer functions and adaptation science for effective decision making.

3. Support the development of research activities as indicated in a recent report from the UK Royal Society on climate change and biodiversity (http://royalsociety.org/). The chief aim of this research would be to improve our understanding of biodiversity in underpinning ecosystem structure and function in climate regulation and in human livelihoods. The interrelationships between biodiversity and climate change require further research and evaluation by the scientific community. In particular, the hypothesis that systems with high biological diversity are more resilient to global change than less diverse systems requires testing.

4. Support the development of scenarios for impacts on biodiversity and ecosystem services under different levels of climate change. Such scenarios are urgently needed now to identify adaptive management priorities and potentially dangerous levels of biodiversity loss.

5. Support the articulation of the benefits, needs and applications of seasonal climate forecasts for adaptation to reduce the effects of climate change and biodiversity losses, in preparation for the World Climate Conference (WCC-3) in Geneva in 2009 (www.wmo.int/pages/world_climate_conferencel/index_en.html).

THE UNITED NATIONS CONVENTION
ON BIOLOGICAL DIVERSITY

www.cbd.intl

In response to this crisis of present and impending loss of biodiversity, the United Nations in 1993 brought into force the United Nations Convention on Biological Diversity (CBD). The three objectives of the CBD are:

1) The conservation of Biodiversity;
2) The sustainable use of Biological Resources; and
3) The fair and equitable sharing of the benefits that result from the use of Genetic Resources.

Scientific studies now make it clear that the climate is changing at regional and global levels and that many ecosystems are already being impacted by these changes. Climate change has been described as one of the major challenges of the twenty-first century to conserving biodiversity, combating desertification and ensuring the sustainable use of natural resources—particularly since the rate of global climate change projected for this century is more rapid than any change that has occurred in the last 10,000 years. Its threats to ecosystems and to the spread of desertification are further compounded by the fact that humans have altered the structure of many of the world's ecosystems through habitat fragmentation, land degradation, pollution and other disturbances, making ecosystems more vulnerable to further changes. Responses to deal with these threats will require improved scientific understanding of the linkages between the climate, biodiversity and the processes of desertification, along with an enhanced environmental forecasting capability to predict potential biodiversity and land-use changes that may occur.

A summary of some anticipated effects of climate change on biodiversity is provided in Table 1. It should be remembered that in a region as vast as the Americas, there are significant differences over short distances and time-scales in the changing climate, its variabilities and extremes. Given the number of scientific studies that point to the differing localized rates of species and ecosystems adapting or maladapting to the changing climate, it is clear that the Americas can ill-afford the loss of even one species.

THE MILLENNIUM ECOSYSTEM ASSESSMENT

http://www.millenniumassessment.org/en/index.aspx

The global Millennium Ecosystem Assessment (MEA) (2005) further clarified the impacts on biological diversity and emphasized that protecting biodiversity is in the self-interest of all humans and their societies. Biological resources are the pillars upon which civilizations are built. Loss of biodiversity threatens essential ecosystem goods and services, while also interfering with the earth's hydrological, weather and climate systems. The various goods and services provided by ecosystems include:

- provision of food, fuel and fiber;
- provision of shelter and building materials;
- purification of air and water;
- detoxification and decomposition of wastes;
- stabilization and moderation of the Earth's climate;
- moderation of floods, droughts, temperature extremes and the forces of wind;
- generation and renewal of soil fertility, including nutrient cycling;
- pollination of plants, including many crops;
- control of pests and diseases;

TABLE A.1. Examples of projected impacts of climate change on biodiversity (adapted from CBD Information Report, Annex 1: Biodiversity and Climate Change, 2007. UNEP/CBO/SBSTTA/12/7)

The Changing Climate	Projected Climate Impacts	Impacts on Biodiversity
Increased Air Temperatures	Increased number of hot days	• Increased heat stress on biodiversity, loss of sensitive species and possible extinctions • Increased exposure to pests and diseases • Increased drying of wetlands and waterways • Invasion by more heat-tolerant species
	Increased water temperature	• Decreased dissolved oxygen • Increase in instances of disease among fish • Loss of cold- and cool-water fish species • Increased vulnerability to invasive alien species • Reduced productivity of marine systems (coral reefs and seagrass beds) and possible extinctions
	Sea level rise	• Salt water intrusion in coastal wetlands and other inland waters (islands especially vulnerable) • Inundation of lowlands and coastal wetlands • Increased mortality and disturbance of critical habitat • Increased erosion (beaches/coastal cliffs)
	Melting permafrost	• Changes in nutrient cycling and soil biodiversity • Reduced access to food sources as a result of repeated freeze-thaw cycles • Loss of cryosoil-based ecosystems and species • Land instability, increased sedimentation and erosion • Drainage of lowland Arctic tundra
	Decreased ice cover (later freeze and earlier breakup)	• Reduced winterkills • Changes in deposition of sediments in floodplains, affecting aquatic life
	Glacial retreat and decreased snow cover	• Changing hydrological regimes • Changes in seasonal cues for mountain biodiversity • Increased predation • Disruptions in hibernation patterns • Reduced insulating protection from snow • Loss of snow bed ecosystems and species

Changes in precipitation regimes	Increased instances of drought	• Loss of ground cover leading to desertification and loss of soil biodiversity • Increased water stress on biological communities • Reduced availability of food and fodder • Salinization, compaction, cementation of soils • Increased risk of fire • Changes in natural flow regimes of rivers and streams • Changes of alpine grassland to steppe
	Increased flooding	• Increased erosion of soil biodiversity • Increased land degradation • Increased threats from waterborne disease • Increased habitat destruction from flooding • Changes in natural flow regimes of rivers and streams
	Decreased freshwater availability in lakes and coastal zones	• Decline of water levels and availability in freshwater lakes • Significant impacts on near shore coastal biodiversity (e.g. bird and aquatic species) • Disappearance of coastal wetlands • Emergence of new land and property ownership issues
Increased frequency of extreme climate events	Disruption in growth and reproduction	• Changes in biodiversity. biomass and productivity • Changes in fires, insects and disease regimes • Increased mortality • Damage to forest structure, alteration of succession patterns and landscapes
	Heightened storm surges	• Increased mortality of ecosystems and disturbance of critical habitat, • Habitat loss (especially mangroves. reefs, sandbars and beaches) • Increased erosion and sediment damage

- maintenance of genetic resources as key inputs to crop varieties and livestock breeds, medicines and other products;
- cultural and aesthetic benefits; and
- ability to adapt to change.

THE UN FRAMEWORK CONVENTION ON CLIMATE CHANGE

http://unfccc.int/2860.php

The UN Framework Convention on Climate Change (UNFCCC) seeks to stabilize greenhouse gas (GHG) concentrations in the atmosphere at a level that will avoid dangerous human interference with the climate system. Because the climate of the future will eventually respond to all of the GHGs collected in the atmosphere over time, even cutting future GHG emissions to zero will not stop most changes. Hence, ecosystems and communities will need to adapt to climate change even if anthropogenic emissions are reduced to near zero.

Climate change is likely to have significant impacts on most or all ecosystems, since the distribution patterns of many species and communities are determined to a large extent by climate. However, ecosystems and biodiversity responses to changes in regional climate are rarely simple. At the most basic level, changing patterns of climate will alter the natural distribution limits for species or biological communities. In some cases, it may be possible for species or communities to migrate in response to changing conditions if there are no significant barriers to migration. Rates of climate change will also be critical and these will vary at regional and even local levels. The maximum rates of spread for some sedentary species, including large tree species, may be slower than the predicted rates of change in climate conditions.

The most vulnerable ecosystems will include those habitats where the first or initial impacts are likely to occur and those where the most serious adverse effects may arise or where the least adaptive capacity exists. These include, for example, Arctic, mountain and island ecosystems. Tools and guidance in the form of scientific predictions of ecological states are essential to pinpoint priority ecosystems and to guide climate change response options.

Organisms and ecosystems have a natural but limited ability to adjust to climate change. It is clear that as the climate has cooled and warmed over the past hundreds of thousands of years, the various major ecotypes and the animal communities that inhabit them have shifted cyclically to the north and south. Projected climate change, primarily driven by human-induced causes, is faster and more profound than anything in the past 40,000 years, and probably the last 100,000 years (IPCC, 2007). The UN Intergovernmental Panel on Climate Change (IPCC) Working Group II report suggests that 20%–30% of global plant and animal species are likely to be at increased risk of extinction if increases in global average temperature exceeds 1.5° to 2.5°C.

THE UN CONVENTION TO COMBAT DESERTIFICATION

http://www.unccd.int/

The UN Convention to Combat Desertification (UNCCD) promotes an innovative approach to managing dryland ecosystems and arid regions, and recognizes that desertification is caused by climate variability and human land management activities. Desertification is defined by the UNCCD as "land degradation in arid, semiarid and dry subhumid areas resulting from various factors, including climatic variations and human activities" (MEA, 2005). Desertification involves the loss of biological and economic productivity, as well as complexity in croplands, pastures and woodlands.

The UNCCD recognizes that combating desertification is necessary to improve conditions in developing countries, particularly the least developed. To combat desertification and mitigate its effects in countries experiencing serious drought and/or desertification, the UNCCD outlines long-term strategies that focus simultaneously on improved productivity of land and the rehabilitation, conservation and sustainable management of land and water resources. Its chief mechanism for implementing them is through the development of action programs to manage dryland ecosystems and arid regions (UNEP, 1996).

Addressing the underlying causes of desertification and drought and identifying measures to prevent and reverse them, action programs have been detailed for Africa, Asia, Latin America and the Caribbean, and the Northern Mediterranean (UNEP, 1996). The UNCCD also recognizes that the implementation of the UNFCCC, the CBD and related environmental conventions will play a significant role in combating desertification.

EXAMPLES OF CLIMATE CHANGE ADAPTATION OPTIONS FOR BIODIVERSITY

Adaptation options can be classified under many general thematic approaches, which include the following:

Adaptation Strategies

- Ensure options are of the do-no-harm variety.
- Improve predictions or plausible scenarios of future climate change.
- Develop National Adaptation Plans that include biodiversity.
- Undertake Adaptive Capacity Studies.
- Prioritize intact or relatively unaffected habitats for protection.
- Engage local people in planning and implementing mitigation and protection strategies.
- Design reserves to protect vulnerable life stages.
- Respond to changes already inherent in the system.
- Improve integrated monitoring and detection programs.

Planned Adaptation

- Build corridors.
- Reintroduce species with great care.
- Assist species regeneration.
- Employ ex situ conservation if extinction is imminent.
- Manage for disturbances to the ecosystem.
- Account for projected effects of climate change when designing new protected areas.
- Track and manage invasives (e.g., control or eradicate invasive species).

Building Ecological Resilience

- Reduce fragmentation.
- Protect space, functional groups, climate refugia and multiple microhabitats in replicated areas.
- Maintain a natural diversity of species, ages, genetic diversity and ecosystem health.
- Provide buffer zones and flexibility of land uses.
- Ensure connectivity of habitats along gradients.
- Reduce other related and cumulative stressors.

Technological Adaptation Solutions

- Efficient management of rain/snow water availability.
- Changes in timing/type of irrigation and fertilization.
- Inoculate with soil biota important to plant vigor.
- Establishment of aquaculture.
- Diversion of fresh water.
- Seawalls, dykes and tidal barriers.
- Bridges to cross inundated areas.
- Increase density and reliability of climate monitoring.

Behavioral Adaptation Solutions

- Early-Warning Climate Alert and Response Programs for Biodiversity.
- Prediction of climate extremes and hazards for emergency preparedness and disaster management of critical biodiversity.
- Risk management assessments and priority behavioral-based action plans to reduce the impacts of a changing climate on the functioning of biodiversity.
- Redefinition of critical biodiversity thresholds to the new multiplier climate.

Regulatory/Policy Adaptation Actions

- Rezone coastal areas.
- Establish protected areas.
- Natural forest regeneration or avoided deforestation.
- Decrease nutrient enhanced run-off.
- Nonchemical control of pest/disease outbreaks.
- Establish no-take zones.
- Landscape scale management of water availability and quality.
- Change trade policies.

Economic Adaptation Approaches

- Changes in grazing management and water management.
- Apply modifications in agricultural land base and incentives for more sustainable agriculture and forestry.
- Offer incentives to control the spread of invasive species.

- Eliminate incentives that accelerate habitat loss.
- Adopt energy efficient technologies for both adaptation and mitigation benefits.

Adaptation Science

- Model the buffering capacity of forest habitat for biodiversity in a changing climate, especially in urban parks and school yards.
- Reduce other pressures/threats.
- Introduce species tolerant to salt, drought, pests or higher temperatures.
- Rehabilitate damaged ecosystems.
- Multi-cropping, mixed farming, low-tillage cropping or low-intensive forestry.
- Apply integrated models for climate and biodiversity prediction.
- Improve the understanding of extremes/hazards and cumulative events.

REFERENCES

[IPCC] Intergovernmental Panel on Climate Change. 2007. *Climate Change 2007: Impacts, Adaptation and Vulnerability.* Contribution of Working Group II to the Fourth Assessment Report of the IPCC. Cambridge, UK: Cambridge University Press.

[MEA] Millennium Ecosystem Assessment. 2005. *Ecosystems and Human Well-Being: Biodiversity Synthesis.* Washington, DC: World Resources Institute.

[UNEP] United Nations Environmental Programme. 1996. *Water and Wastewater Reuse: An Environmentally Sound Approach for Sustainable Urban Water Management.* Osaka, Japan: UNEP, Division of Technology, Industry and Economics, International Environmental Technology Centre.

Index

About the Contributors

Federico E. Alice is a forestry Engineer who graduated from the National University of Costa Rica, where he has worked for the past five years as a researcher at the Forest Research and Services Institute (INISEFOR). During this time he has taken part in research aimed at the restoration of degraded lands through forestry activities such as native tree plantations and natural regeneration of forests. Key elements of this research were biological carbon sequestration and storage, financial viability of land use change (from non forest to forest) and the carbon markets. At the same time he has worked as a consultant along with project developers in the different stages of the carbon project cycle, especially for projects pretending to participate from the voluntary carbon markets. Most of them have been related to the forestry sector (afforestation and reforestation (A/R) and reduced emissions from deforestation and degradation (REDD) activities). Current work interests: emissions trading, carbon markets, offset project protocols and standards.

Holly Alpert received her B.A. from Wellesley College, and her Ph.D. in Environmental Studies at the University of California, Santa Cruz. Her interdisciplinary dissertation focused on the implications of climate change for conifer establishment and water resources management. She currently works to use natural science to inform environmental policy.

Fred L. Bunnell studied forestry and wildlife biology in Canada, Switzerland and the United States—including Berkeley in the 1960s, which did not leave him unchanged. He has been at the University of British Columbia for more than 30 years where he was Professor of Forest Wildlife, Forest Renewal BC Chair in Conservation Biology, and founding Director for the Centre of Applied Conservation Biology. Bunnell is now a Professor Emeritus in the Center for Applied Conservation Research. He has published over 400 scientific articles and reports to government and industry, received 6 national and international awards for applied research, held commissions and served on over 70 provincial, national, and international committees dealing with resource management. His work has taken him to 30 countries. He is happily married, a father, and enjoys life immensely.

Sonia B. Canavelli is a wildlife biologist at the National Institute of Agricultural Technology (INTA) in Argentina. Ten years ago she was involved in the development of protocols to evaluate the impact of pesticides on wildlife in agroecosystems at the country level and the coordination of fieldwork projects related to Swainson's hawk ecology in Argentina, a migratory bird affected by the misuse of pesticides in the country. Together with María Elena Zaccagnini, she designed and coordinated the bird monitoring program between 2001 and 2003, as a way of evaluating the status of bird populations at a regional scale. At present, she collaborates with the implementation of the monitoring program, data organization and analyses. Also, she is coordinating an INTA national project on the ecology and management of conflicts between wildlife and agricultural production. She received a B.S. in Biology from the University of Córdoba, Argentina, and an M.S. in Wildlife Ecology and Conservation from the University of Florida, Gainesville, where she is a Ph.D. candidate.

Noelia C. Calamari is a biologist at the National Institute of Agricultural Technology (INTA) Argentina. Since 2004, she manages the operacional and technical aspects of the bird regional monitoring program in the central argentine pampas agroecosystems. She organizes the data base and analyzes data for estimating bird abundance and richness. She also develops maps and spatially explicit models of bird distribution in agricultural landscapes and forests. Nowadays, her main research interest focuses on bird responses to forest fragmentation thresholds in the Espinal, Chaco forest subecoregion. She annually trains bird observers for field sampling and cooperate in the design of sampling schemes for biodiversity monitoring in different ecoregions. Noelia has her Bachelor in Biology degree from the Universidad Nacional del Litoral. In 2006 she graduated from Universidad Nacional de Lujan as Specialist in Teledetection and Geografical Information Systems applied to natural resource management. She is a Ph.D. candidate in the Biology School of the Universidad Nacional de Córdoba, Argentina.

Francisco Dallmeier is the Director of the Center for Conservation Education and Sustainability (CCES) of the Smithsonian's National Zoological Park. His main work focuses on integrating biodiversity conservation into sustainable development. He has coordinated the Smithsonian Monitoring and Assessment of Biodiversity Program (SI/MAB) since 1986 to promote research and long-term monitoring for over 300 forest plots worldwide. He led the nomination of the Smithsonian Conservation and Research Center as core site of the National Science Foundation's National Ecological Observatory Network, NEON. He is leading the Conservation Studies Partnership with George Mason University and is the coprincipal investigator for the Smithsonian Energy and Biodiversity programs in Peru and Gabon, and the Smithsonian GlobalTiger Conservation Initiative with the World Bank. He is author, coauthor or editor of more than 130 publications. Dallmeier earned his B.S. in biology from the Central University of Venezuela and his M.S. and Ph.D. degrees in wildlife biology, from Colorado State University.

Anthea Farr divides her time between activities as a consulting biologist and volunteer. As a consulting biologist her work is more clearly natural history than specialized, and has addressed a wide range of organism groups from plants through animals. Much of her recent effort has been directed to increasing awareness of elected officials and civil servants about urban opportunities in conservation. A large portion of her volunteer work represents her answer to Richard Louv's *Last Child in the Woods* and providing opportunities for youth to learn and enjoy nature through the "Young Naturalists." Her work with the "Young Naturalists" has been recognized by the Daphne Solecki Award of the BC Federation of Naturalists, but she would continue the work without that recognition.

Adam Fenech is a Ph.D. Climatologist who focused on the rapid assessment of the impacts of climate change at Physical and Environmental Sciences, University of Toronto. He has extensive experience in teaching, training and writing in the areas of linking global climate models to local decision-making, and has been a research collaborator with the Smithsonian Institution for over 15 years. He is the Associate Director of the Adaptation and Impacts Research Division of Environment Canada, having worked in various positions with Environment for over 22 years.

William Fonseca is a Forestry Engineer with an M.B.A. in Business Administration and as a Ph.D. candidate he has completed the Global Change and Sustainable Development Program from the Universidad de Alcalá de Henares, Spain. He obtained his first degree in 1983 and since then has worked in research and as a professor at the National University of Costa Rica. Has participated in different research projects on subjects such as fertilization in forestry plantations and nurseries, growth evaluation of native and nonnative forestry plantations, growth models, thinning trials in forestry plantations, natural forest management, and for the past five years, has worked on biological carbon sequestration and storage. These works have allowed him to publish more than 60 papers in different national and international journals. He has developed a similar amount of consultancies on a national and international level and has also participated in more than 30 seminaries, workshops and congresses where he has been able to share his work experiences.

Oscar R. García-Rubio received a B.S. in Biology and a M.S. in Biochemistry Sciences from Universidad Nacional Autónoma de México. He completed his Ph.D. in the Universidad Autónoma de Querétaro. His research interests and publications are varied, including topics related to plant bioenergetics, physiology and plant reintroduction. Of special interest have been effects of climatic change on plants and microorganisms assembling in arid and semiarid ecosystems. In 2008 he received the Alejandrina Award in Basic Science and Environment.

Alden B. Griffith received his B.A. from Wesleyan University and a Ph.D. in Environmental Studies from the University of California, Santa Cruz. His research interests include impacts and interactions of non-native plant species, ecological effects of climate change and plant-plant facilitation. He is currently the postdoctoral Botany Fellow at Wellesley College.

Chrystal Healy received her B.Sc. and M.Sc from the University of McGill, Montreal, where she worked on the Sardinilla Project exploring the role of tree biodiversity and ecosystem function. She has since moved toward the realm of corporate sustainability and is now currently the Manager of Environmental Affairs for Quebecor Media.

Marianne Karsh is a researcher for the Adaptation and Impacts Research Division at Environment Canada. She is the author/coauthor of various scientific publications, including the keynote paper "Climate Change and Biodiversity in the Americas," Environment Canada; "Climate Change and Biodiversity: Implications for Monitoring, Science and Adaptive Planning," Environment Canada; the *Adaptation Science* newsletter and Occasional Paper Series, Environment Canada; *Climate Change and Biodiversity: Perspectives and Mitigation Strategies*, the Icfai University Press, India, and the Canadian Forest Service Information Report Series. Marianne was a member of the Organizing Committee and presenter at the Climate Change and Biodiversity in the Americas Symposium, Panama, February 2008. She has considerable experience as an invited speaker to various groups both nationally and internationally. Marianne is the Director of Arborvitae, an organization whose mission is to advance ecological awareness, education and commitment with adults, children and youth. She is also the coordinator of the Ecology Project for the Ignatius Jesuit Centre overseeing a number of initiatives combining ecological activism and justice. Marianne has worked as a Forest Mensurationist at the Ontario Ministry of Natural Resources and as a Researcher for the Canadian Forest Service before starting her own organization, Arborvitae and then joining the team of researchers at Environment Canada.

Humberto Leblanc was born October 25, 1960, in Panama. After attending public schools in Panama, he received the following degrees: B.S. in Agronomy from the University of Panama at Panama (1984); M.S. in Crop Systems and Plant Breeding from the Tropical Agricultural Research and Higher Education Center (CATIE) at Costa Rica (1991); M.B.A. in Business Administration from the Central American Autonomous University (UACA) at Costa Rica (1999); and Ph.D. in Agronomy from the University of Missouri-Columbia (2004). He has been a professor at EARTH University at Costa Rica since 1993.

Michael E. Loik received his B.Sc. and M.Sc. from the University of Toronto, and Ph.D. at the University of California, Los Angeles. His research interests focus on the impacts of changing rain and snowfall patterns on ecosystem patterns and processes. He is currently Associate Professor of Environmental Studies at University of California, Santa Cruz.

Don MacIver is the Director of the Adaptation and Impacts Research Division at Environment Canada. He has a long and distinguished research record with over 100 scientific publications to his credit and the Nobel Peace Prize. Don has considerable international experience having represented Canada at numerous scientific meetings of the International Panel on Climate Change (IPCC), UNFCCC Conference of the Parties, Inter-American Institute for Global Change Research (IAI), World Meteorological Organization (e.g., Co-Chair of the Organizing Committee for the World Climate Conference—3), UNESCO's Man and the Biosphere Programme and the UN Convention on Biological Diversity. He has been a Research Collaborator of the Smithsonian Institution. Don has served as a Professor at York University and as an adjunct Professor at the University of Toronto. He worked as a Forest Biometrician and Mensurationist at the Ontario Ministry of Natural Resources before joining Environment Canada as a forest meteorologist and adaptation scientist. Outside of work, he is a Municipal Politician (Mayor) and County Councillor, a farmer and an environmental activist.

Guadalupe Malda-Barrera is a Professor at the Universidad Autónoma de Querétaro. She received her B.S. in Biology in the Universidad Autónoma Metropolitana, Mexico, and her

Ph.D. in Botany at Arizona State University, USA. She was a research scientist at Instituto de Ecología y Alimentos at Universidad Autónoma de Tamaulipas for 13 years before joining the Universidad Autónoma de Querétaro. Her research interests focus on the in vitro culture of native plants and their hardening, as well as the ecophysiology of in vitro cultured cacti. Malda co-organized the first Symposium of Biosfera Reserve of "El Cielo" in 1988. At the present time she is Coordinator of Biology College at Universidad Autónoma de Querétaro. In 2008 she received the Alejandrina Award in Basic Science and Environment.

Johan Montero Eduarte graduated from the National University of Costa Rica (UNA) as a forestry engineer in 2005. From 2005 to 2007 he worked at the Forest Research and Services Institute (INISEFOR) of the National University. In 2007 he started labors at the Instituto Costarricense de Electricidad (ICE), where for 2 ½ years worked at a watershed management unit in San Carlos. There, he worked on subjects such as a program on payment for environmental services (PSA), nursery production of forest plants, as well as proposing a watershed management plan for the area. Nowadays, he is improving his knowledge in the area of land valuation at ICE.

Catherine Jeanne Potvin is currently a Full Professor in the Department of Biology at McGill University, Montréal, Québec, Canada. She received her B.Sc. and M.Sc. from the Université de Montréal and her Ph.D. from Duke University. During her career she has worked on issues related to global climate change. In the recent years she have developed an expertise on tropical forest ecology and carbon storage. This led her to establish a research platform, the Sardinilla Project, where she is looking at the role of tree biodiversity as a determinant of ecosystem function. Catherine has also been working with the Embera people of Panama since 1994 and developed participatory approaches to integrate the human dimension in biological analysis. Since 2005, she has worked closely with Panama's Environmental Authority as their Special advisor on forests and carbon. She has published 61 papers in Refereed Journals, referred 12 chapters in Books, edited 2 books and written 5 articles for the general public.

Michael Preston is a Director and cofounder of the Biodiversity Centre for Wildlife Studies, Manager of the Wildlife Data Centre and Editor of Wildlife Afield. He also was a Research Associate working with Dr. Bunnell at the University of British Columbia at the time of writing the paper in these proceedings. He is now supervising and enacting programs within Jacques Whitford, an environmental consulting firm of 1,700 employees now part of Stantec. Prior to assuming his current position, Preston's consultancies included BC Hydro, Burrard Inlet Environmental Action Program, Canadian Forest Products, Canadian Wildlife Service, Coast Salish Nations, Crestbrook Forest Industries, Tembec Inc., and Weyerhaeuser. He has authored over 80 popular articles, technical reports, and reviewed papers.

Malena Sarlo currently works at the Nature Conservancy as a Conservation Planner. She provides technical and scientific support for conservation initiatives in planning and monitoring schemes. She performs a variety of scientific tasks and/or conservation activities, especially for the landscapes where TNC works in Panama, namely Darien, Amistad-Bocas del Toro, Western Pacific Coast of Panama and Chagres. Apart from Panama, she is now in transition to work in conservation activities in Costa Rica, Colombia, Venezuela and Ecuador.

Anne Schrag is currently the Network Coordinator for WWF's U.S. Climate Adaptation Network. In this position, she works to coordinate climate adaptation efforts in the Northern Great Plains and across the country and works directly with state and federal agencies to "climate proof" their management plans. She has worked on climate-change issues—from both an ecosystem and large-scale planning point of view—for the past seven years. She recently completed a study on the impacts of climate change on greater sage-grouse habitat in the Northern Great Plains, which is the first of its kind in regards to spatial resolution of future climate data used in the study region. She has also published research on the impacts of climate change in the Greater Yellowstone Ecosystem and the Pampas region of Argentina. For the past seven years, she has used spatial-analysis techniques to help understand climate-change impacts on ecosystems of conservation concern. Prior to working with WWF, Anne served as Program Coordinator for the National Park Service Greater Yellowstone Inventory and Monitoring Network, where she assisted in the development of indicators for monitoring ecosystem health in the national parks of the Greater Yellowstone Ecosystem. She received a BS in Ecology and Evolutionary Biology and a BA in Spanish Literature from the University of Kansas in 2001 and completed her MS in Land Resources and Environmental Sciences at Montana State University in 2006.

Robert C. Szaro is currently retired. He was Chief Scientist for Biology for the U.S. Geological Survey in Reston, Virginia from August 2004 to July 2008. From July 2000 to July 2004 he served as Deputy Station Director for the U.S.D.A. Forest Service's Pacific Northwest Research Station in Portland, Oregon, USA. Previously, he served as Coordinator for the Special Programme for Developing Countries of the International Union of Forestry Research Organizations (IUFRO-SPDC) and the Agricultural Attaché (Forestry) for the U.S. Embassy in Vienna, Austria (August 1996 to June 2000). In his capacity as Coordinator for IUFRO-SPDC, he was responsible for forestry research capacity building efforts throughout the developing world with particular focus primarily on Africa, Asia and Eastern Europe. From 1989 to 1996, he served in several capacities in the Forest Service's National Headquarters in Washington, D.C., including Ecosystem Research and Biodiversity Specialist, Research Budget Coordinator, Special Assistant to the Deputy Chief for Research on Ecosystem Management and Special Assistant to the Chief on the "Ecological Stewardship Project." He has authored more than 120 papers and edited 3 books on the conservation of biodiversity, sustainable resource management and the implementation of ecosystem management. Szaro received a B.S. in Wildlife and Fisheries Biology from Texas A&M University (1970), a M.S. in Zoology from the University of Florida (1972), and a Ph.D. in Ecology from Northern Arizona University (1976). He also completed the Senior Executive Fellows program at Harvard University (1993).

Henry Toruño graduated as a Forestry Engineer in 1998 from the National University of Costa Rica and as an M.Sc. in Environmental Management in 2008. He worked for eight years in the Soil Laboratory at the Forestry Services and Research Institute (INISEFOR) from the National University of Costa Rica and collaborated with the different research activities conducted at this Institute. He is now based at Nicoya, Guanacaste, where he works for the Mesoamerican Centre for Sustainable Development in the Dry Tropics (CEMEDE), also from the National University. There he conducts research on water management, agroforestry systems and environmental education.

Adrianne G. Tossas earned a Ph.D. degree in ecology from the University of Puerto Rico, Río Piedras campus in 2002. As her dissertation topic she studied the breeding biology, distribution in the landscape and metapopulation structure of the endemic Puerto Rican Vireo (*Vireo latimeri*) in southwestern Puerto Rico. She also described the effects of the passage of Hurricane Georges in 1998 to the avian community in her study site. From 2002 to 2004 she implemented the Important Bird Areas program of BirdLife International in Puerto Rico, and since then have taught and conducted research at higher education institutions. Adrianne collaborates in projects of the Society for the Conservation and Study of the Caribbean Birds and the Puerto Rican Ornithological Society, particularly promoting long-term monitoring programs and increasing public awareness on the need to protect the region's unique avifauna. As part of these efforts she founded the Puerto Rican Shorebird Network in 2001 to record species richness and abundance in 14 locations throughout the island and the Caribbean Endemic Bird Festival in 2002 as a tool to educate Caribbean nationals about avian conservation. This festival is celebrated annually from 22 April to 22 May in 13 islands throughout the region. At present, Adrianne is a biology professor at the University of Puerto Rico, Aguadilla campus, where she also mentors undergraduate students conducting their first research projects.

Prof. María E. Zaccagnini is the Conservation and Sustainable Use of Biodiversity in Agroecosystems Research Leader for the National Institute for Agriculture and Livestock Technology, Professor of Landscape Ecology at the Autonomous University of Entre Ríos (both in Argentina). In this position, she works to coordinate a research network focused on biodiversity monitoring and sustainable use to understand how patterns of agricultural intensification and expansion affect ecological processes and provision of biodiversity ecosystems services. Known for her leadership in the development of the Wildlife Ecology and Management Sub-Program at INTA, (period 1991–1999), her work has helped to change the mentality of the productive oriented institution into more ecologically oriented research agriculture. Under her coordination, INTA developed a research line on integrated pest management of bird pests, and also took control of the monitoring and remediation of the impact of pesticides on migratory and resident birds. She designed a program and methodologists for monitoring impacts of pesticides that is followed as a model in many Latin American Countries. She and INTA received the 1997 Special Conservation Award from U.S. Fish and Wildlife Service for her contribution to conservation of Swainson Hawks in their wintering grounds. In 2001, the Scientific Argentine Community recognized her contributions for conservation in agroecosystems with the "Francisco de Asis" Award. As a Professor at the Autonomous University of Entre Ríos, she is introducing the concepts of Landscape Ecology in a traditional Ecology Program, and innovative theoretical and practical approaches. María Elena volunteered the IUCN-SSC-Sustainable Use Specialist Group, helping the development of analytic work for assessing sustainability of uses of biodiversity, and shared the responsibility for planning the international agenda on sustainable use issues. She received a BS in Teaching Biology at Universidad Nacional del Litoral in 1976, and completed her MS in Fisheries and Wildlife Biology at Colorado State University in 1989.